A Surveyor's Lot

A Surveyor's Lot

John Smallwood

To Pam
John Smallwood

NORTH
Publishing

First published in December 2000 by
North Publishing
5 Station Road
Cullingworth
Bradford BD13 5HN

© John Smallwood 2000

All right reserved. No part of this book may be reproduced, stored in a retrieval system, or transmitted in any form or by any means, without the prior permission of the author.

Maps are reproduced from the Ordnance Survey map with the permission of Her Majesty's Stationery Office © Crown Copyright, NC/00/1208

ISBN 0-9539828-0-7

Editorial Assistant Angela Holmes
Typeset in Bembo by Pauline Ainley
Cover design by Tony Ainley

Printed in England by Edward Mortimer Ltd
Pellon Lane Halifax HX1 4AD

Dedicated to
my late brother Douglas
for all the good work he did
during his lifetime with
Dacre, Son & Hartley

Preface

It was in 1989 that Dacre, Son & Hartley sold their goodwill to the Abbey National Building Society. This caused me, at the age of 75, to write *An Auctioneer's Lot* in 1990. In this I recorded that only time would tell if it was a good thing for the Society or the old firm. I added that people who knew me would understand there was no way I would ever have sold my independence, no matter what financial temptation was offered.

Apparently the transaction didn't work for Dacre, Son & Hartley, for after a short while they re-purchased the goodwill and are again the old firm which I and many others loved so much.

An Auctioneer's Lot surprised me as it sold so well. I have received letters about it from America, Australia, Denmark, and the Canary Isles. So, after another ten years and at the age of 85, I have been tempted to write this, *A Surveyor's Lot*. Should it be as well received as my earlier effort, I could in another ten years be publishing *That's My Lot!*

John Smallwood

December 2000

Contents

CHAPTER		PAGE
1	Early days	1
2	Wartime	22
3	Post-war	29
4	Clearances	36
5	Mosque	41
6	Wrongful dismissal	45
7	East Hunslet Liberal Club	48
8	Swine Lane, Keighley	52
9	Matthew Horton's opinion	58
10	Ransom strips	62
11	Outline planning application	68
12	Plans rejected	78
13	Public Inquiry	86
14	Financial negotiations	107
15	The end of the saga	110
16	Memories	114
17	Other professions	116

The author aged 8

CHAPTER ONE

Early days

As you grow old there are some things you never forget. Life started for me in 1915 as a war baby and I lived with my father, mother, sister and brother at Sandal, a residential suburb of Wakefield.

At that time the buses had not reached Wakefield but there were trams that ran from Sandal to Leeds taking an hour and costing one shilling and three pence return fare. The Wakefield trams were painted green, with no windscreen to protect the driver and open platforms upstairs at the front and back, each with a bench. The buses arrived in 1920, single decker Bristols with solid tyres.

The railway system at that time was owned by public companies with many shareholders. Yorkshire was served by the London & North Eastern Railway Co, the LNER, which ran into Kings Cross and the London Midland and Scottish, the LMS, which ran into Euston and St Pancras. Other companies were the Great Western serving the West Country and running into Paddington and the Southern Railway which ran from the south coast into Victoria. The Railway Companies acquired the right under various Acts of Parliament to buy land for the railways from landowners and compensation was fixed under the Acquisition of Land Acts. Landowners receiving compensation from land acquired by Utility Companies such as the Railways had the compensation fixed under the Land Clauses Act which entitled the claimant to add a bonus of 10% over and above the compensation which would have to be paid by a local authority.

There were some unusual arrangements when the Railway Companies developed. The LNER Leeds to London line passed over land purchased from Sir Thomas Pilkington at Sandal. In the sale of his land to LNER it was agreed in the deal that anybody living on land owned by Sir Thomas could, by arriving at Sandal

Station fifteen minutes beforehand, stop any train and board it. My father exercised this right many times but the privilege was lost when the Railways were nationalised. Travel was by First Class or Third Class and I never did find out why there was no Second Class, with the exception of the Southern Railway.

My father and mother took the family to the Wembley Empire Exhibition in 1923, staying at the Great Northern Hotel at Kings Cross. Private facilities were provided in the way of a washstand with pottery bowl and jug, and by ordering the night before, a jug of hot water was delivered to the bedroom door for washing and shaving needs. China chamber pots were provided in little bedside cupboards and when the chambermaids called, they emptied them along with the toilet bowls into slop pails. A few years back I saw one of these slop pails in use as a punch bowl at a party being given by Jim Feather, one of my partners. All present were too young to realise the punch bowl's previous use; I stuck to lager.

I was sent as a boarder to Southcliffe School, Filey at the age of eight in 1923 with fees of £50 a term; the staff were underpaid including a very good, enterprising, but unfortunate Art Master who received seven years for printing his own bank-notes. In 1938 at the age of thirteen I passed my Common Entrance Exam (23%) at the height of the slump and joined 365 other boarders, all boys, at Wrekin College, Wellington where fees were £125 a term. In those days maids were not costly, we had our beds made for us and also waitress service in the dining room, but were not allowed to talk to them. It was probably thought that if we did we could become excited. The only female teacher employed taught music. Once a year we attended the local operatic annual presentation and the local girls' school attended the same evening. We were separated from the girls by a rope.

At Wrekin there was no electricity but we had gas, rather like Tom Brown's day. It was very efficient and was switched on at the door in the same way as electricity today. The educational aim was to pass Oxford and Cambridge Matriculation, this was before the birth of the school certificate. I found myself in a class called Middle 5C Matric which comprised in the main, boys who were going to enter their father's business and had no intention of matriculating. The school at the time had embarked upon refurbishing, building a new science block and making general improvements. There was shortage of space and it seemed natural for Middle 5C Matric to be allocated the woodwork shop for lessons for the duration of a term, the heating being provided by a gas fire. This also coincided with the arrival of a new school Chaplain the Rev Nash, being of the type unsuited to teaching, not even scripture.

One winter's morning a classmate, one of three South American brothers boarding at the school, called Echivarer, had obtained a blank cartridge from the school cadet range and before Rev Nash entered the workshop Echi removed a porcelain strip from the gas fire, inserted the cartridge and replaced the strip. Rev Nash started the lesson saying he was going to give us a scripture test; up shot Echi's hand "Please Sir, its very cold can we please have the fire lit." Rev Nash got out his matches, bent down and lit the fire and we boys ducked down behind our bibles for protection. He started to write questions on the blackboard and after about a minute there was a terrific explosion, the desk fronts were embedded with porcelain from the fire and Rev Nash leapt in the air with shock leaving a chalk mark on the blackboard about a foot from where he had been writing. Without a smile on his face Echi held up his hand and said "Please Sir, the gas fire's blown back." I do not know what Echi is doing now but he should have gone far. One of his side-lines was to run a book on the length of Rev Nash's Sunday morning sermons.

As a scholar of sixteen I went to London to sit the Law Prelim: I passed in Latin but failed in English, not many could do that. In 1932 at the age of seventeen I left Wrekin and joined Hollis & Webb, Chartered Surveyors at Leeds as an articled clerk. My father paid the sum of 300 guineas for the privilege. I received no wage and office hours were 9 am to 6 pm weekdays and 9 am to 1 pm Saturdays. There was no office car, we did our work on foot, bus or tram; when we wished to impress the client a taxi was used.

The first job I remember was the Probate Valuation of the furnishings at Castle Howard when the Hon Jeffrey Howard died. My job was to write down the value in code as the senior partner dictated to me, the reason for the code being that if anyone was listening they would not know the figures. The job took fourteen days and I lodged with an old lady on the Estate, a temperance worker which I found a bit of a handicap.

In my exams I spent three months each year in London on a revision course at the College of Estate Management and I worked very hard but enjoyed it. I stayed at the West Central Hotel, Southampton Row, since demolished and now the site of the New Bedford Hotel. I paid £4 0s. 6d. a week full board. One thing in London I missed was green fields and I often went out into the country on a Saturday just to get out of the city. I think probably the subject I liked best was the Law of Compulsory Purchase and the Compensation going therewith. We were fortunate to have as our Tutor on the subject, David Lawrence, a first class man who had written several books and was an authority on the subject. I was twenty years old and he took me on one side and told me that he had noticed my work and asked if I would be interested in a job at the College correcting papers returned by the Correspondence Course students. This would not be difficult as all I would have to do was to cross out where the answers were wrong or could be improved, and attach a standard correction prepared by himself. The offer did not appeal to me but I felt honoured when David Lawrence proposed me as an Associate of the Institute of Arbitrators with Compulsory Purchase of Property being my specialist subject.

In March 1936 at the age of twenty, I passed my final examination in both the Chartered Surveyors Institution and the Auctioneers Institute.

Leeds City Police,
"B" Division
Superintendents Office,
Meadow Lane Police Station
Leeds 21st. November, 19 34
11.

TELEPHONE NO. 20301 (7 LINES)
FIRE STATION 20222.

R.L. MATTHEWS, C.B.E.,
CHIEF CONSTABLE.

PLEASE QUOTE
OUR REF. IR/2421-B.
YOUR REF.

ALL COMMUNICATIONS TO BE ADDRESSED "THE SUPERINTENDENT." AND NOT TO THE UNDERSIGNED BY NAME.

Sir,
    I am directed by the Chief Constable to say that it has been reported to him that:

You drove motor car 'U.G.332' on a road and used the motor horn during prohibited period in Thwaite Gate in this City, at 12-15 a.m. on Sunday, 18th. November,1934.

    After having given this matter his careful consideration, the Chief Constable has decided that he will, in the present instance, refrain from taking further steps. He will, however, feel bound to take the circumstances into account if you should again be reported for an alleged offence.

              I am, Sir,
              Your obedient Servant,

              Herbt Hunt
              Chief Inspector
          for Superintendent.

REGISTERED.

Mr. John North Smallwood,
14, Carr Lane, Sandal,
Wakefield.

In 1934 I received an interesting letter from Leeds City Police: I do not think a Chief Inspector, when writing to a nineteen year old today would sign as 'Your obedient servant'.

THE YORKSHIRE TRACTION COMPANY, LIMITED.

TELEPHONES
TELEGRAMS
BARNSLEY
776 & 777

ALL COMMUNICATIONS TO BE ADDRESSED TO THE COMPANY AND NOT TO INDIVIDUALS

UPPER SHEFFIELD ROAD,
BARNSLEY, YORKS.

YOUR REF
OUR REF. AW/MC.

December 11th 1934.

:: MOTOR ::
OMNIBUS
SERVICES
PARCELS DELIVERY
AND
GOODS TRANSPORT

GENERAL MANAGER:
G. W. ROBINSON

REGISTERED OFFICE:
88 KINGSWAY
LONDON. W.C.2
PHONE: HOLBORN 7888

OFFICES & GARAGES:
BARNSLEY
CHIEF OFFICES | PHONE
AND GARAGES | 776 & 777
BURLINGTON ARCADE | 86
(PARCELS & BOOKING)

DONCASTER
30 WATERDALE
(PARCELS & BOOKING)
MILETHORNE LANE | 1332
(GARAGE)

HUDDERSFIELD
7 LORD STREET
(PARCELS & BOOKING) | 3389
ST. ANDREW'S ROAD | 2718
(GARAGE)

ROYSTON, YORKS.
GARAGE   9

WOMBWELL
OFFICE & GARAGE 67

Mr. A Smallwood,
14, Carr Lane,
Sandal,
Nr. Wakefield.

Dear Sir,

We shall be pleased to receive from you per return, the sum of 4½d., 3d of which represents the fare from Wakefield to Carr Lane, when our Conductor allowed you to travel and you had no money, the balance being the cost of postage of this letter.

Yours faithfully,

p.p. THE YORKSHIRE TRACTION COMPANY LTD.

In December 1934 my father received a letter from the Yorkshire Traction Bus Company. Those were the days!

Dacre & Son, Otley Office 1932

In 1932 my brother Douglas had purchased Dacre & Son of Otley for £400, the price included the freehold office, goodwill, fixtures and fittings. Early in 1936 Dacre and Son which had been established in 1820 merged with Douglas Hartley of Ilkley and on 1st April (an appropriate date) I opened an office for the new firm Dacre, Son and Hartley in Skipton at a rent of fifteen shillings a week. My terms were that I received no wage, I could keep what I made and as soon as I achieved £800 profit in a year I could be an equal partner. I did this in my first year.

I had been in Skipton only six months when I received an invitation from Sir Montague Burton for lunch on the opening of their new establishment in the town. I wondered why I had been asked, but it was good for me as I met the notoriety of the town; no doubt they wondered who I was.

Before opening the office in Skipton I had no knowledge of the

> Sir Montague Burton
> requests the pleasure of the Company of
> J. N. Smallwood Esq
> at a Luncheon to be held on Friday, October 9th, 1936
>
> AT THE BLACK HORSE HOTEL, AT 12-30 P.M.
> TO CELEBRATE THE OPENING OF THEIR NEW ESTABLISHMENT,
> AND TO ATTEND THE OFFICIAL OPENING AT 12-0 NOON.
> CAROLINE SQUARE (CORNER OF KEIGHLEY ROAD), SKIPTON
> THE CHAIRMAN OF THE U.D.C., COUNCILLOR M. H. MORRIS, J.P.,
> WILL PERFORM THE OFFICIAL OPENING, AND BE THE PRINCIPAL GUEST AT THE LUNCHEON.
>
> R.S.V.P. BY RETURN TO
> MR. A. W. ROBERTS
> HUDSON ROAD MILLS, LEEDS, 9.

town but I seemed to be accepted by the locals. The work was interesting but did not always go to plan. One of my first instructions was from a lady who was a gent's hairdresser in the High Street. She considered her rateable value was too high. I duly appealed, obtained a small reduction and informed her. Before I could send a bill I received a letter from the client thanking me and also enclosing 50 Craven A cigarettes, adding that she assumed I would rather be paid in kind thus avoiding income tax.

Another early instruction was from a solicitor named Eustace Vant in the nearby town of Settle. He instructed me in an appeal against the rates on a housing estate of some twelve houses and having prepared my valuation I was somewhat surprised when Mr Vant told me he was going to appear and call me as a witness. Usually it is a matter of value and the surveyor conducts the appeal without the help or hindrance of a solicitor. Being young I did not demure. Mr Vant then told me that there were two particular houses he wished me to use as comparisons and I did not appreciate his reasons at that point.

The following day the Assessment Committee dealt with the appeal. I expected Mr Vant would conduct the appeal and call me to give evidence of value. He stood up and said that he was acting for the ratepayers and he therefore called me to deal with the valuations and then sat down. I was not cross-examined by Mr Vant but only by the Local Authority Valuers. I did obtain a small reduction but it taught me at an early age to be prepared for the unexpected. It was not until after the hearing that the Council Valuer told me that the two houses Mr Vant had asked me to treat as comparables were the private residences of the Town Clerk and the Council Treasurer.

My life was wonderful in Skipton: I lodged with the Gills, a farming family at Barden at a charge of £2 per week for bed, breakfast and an evening meal. I had a contract with Wendy's Cafe in the High Street for lunch and shared a table with Wilfred Fattorini who was in jewellery and mail order, Stephen Brown a young solicitor who later became the Coroner, and Lloyd Rolfe, the Managing Director of Aspinall's Paint. I learnt a great deal from Wilfred Fattorini but two things particularly I have remembered all my life:

> 1 Pay all your bills promptly. If you do not, you cannot complain if people are slow to pay you.

> 2 When negotiating with a civil servant remember the best day to do so is on a Friday just before his weekend break and avoid negotiating on a Monday when he is not likely to be in his most generous mood.

The first job I did in London was for Wilfred Fattorini: I travelled down to St Pancras in the morning and did the job in the afternoon, travelling back in the evening. My fee was £5 (train fare £2, job £3). They were great days.

I played rugby with Skipton and it was whilst doing so that I met half-back Jack Leach, a young solicitor whose father Bill Leach was the landlord of the Devonshire Hotel in Skipton. Jack Leach had been articled to Turner & Wall in Keighley and when he qualified as a solicitor they had offered to employ him at £2 per week. This prompted Bill Leach to pay a visit to Harry Wall the senior partner, to tell him that his son was now a solicitor and £2 a week was an insult, he could earn as much as that if he was a butcher's boy. This brought the question from Harry Wall as to why his son had not gone into butchering if the wage was so good. Jack Leach remained with Turner & Wall and in a short time became a partner when I assumed he would earn more than £2 per week.

It was Jack Leach who persuaded me to open an office in Keighley, with a population of 50,000 and not a chartered surveyor in town. I paid a visit to the town and looked for an office. While walking up Devonshire Street I noticed a vacant property at No 24 and negotiated a three year agreement at a rent of 15 shillings a week. The landlord, Joe Scott also wrote in an option for me to purchase the property for £1,000 which I exercised during the last week of the tenancy. I opened the Keighley office on 1st April 1938 and Bill Varley from the Otley office went to Skipton to replace me. Keighley was a very different town to Skipton but just as good a life for me. I obtained an interesting list of important dates from 1780 to 1935 (opposite).

Dacres as we were called for short, produced a property guide called *The Wharfedale, Airedale and Craven Property Register.*

It was a comprehensive list of properties for sale and to let. In a very short time we had become the leading Estate Agents in town and as we were the only Chartered Surveyors in the district, we handled more than our fair share of the commercial work. We were a very broad practice, both residential and commercial. Keighley had everything and our sales included properties of all sizes, 'back to backs' as well as lovely detached residences. Building

1780	June 30	Low Mills Keighley. The First Cotton Mill in England commenced running.
1808		Keighley contained 786 Houses 13 of which were Public Houses.
1824		Population of Keighley 7000.
1850		Amount of Rates collected £3295.5s.9½d
1861		Population of Keighley 21859.
1862	April 1st	*Keighley News* first published.
1864	February 13th	Green House Farm Murder.
1864	October 1st	Earthquake shock in Keighley.
1867	April 13th	Worth Valley Railway opened.
1875	April 15th	Jim Jeffries born, became Heavy Weight Champion of the World after beating Bob Fitzsimmons.
1876	March	Keighley Cottage Hospital opened.
	March 4th	Keighley Baths and Wash Houses opened.
	December 6th	Thwaites Gas Works opened.
1878		Charlie Peace caught at Blackheath by P.C. Robinson. Executed Armley Leeds 1879
1880	December 24th	Rodney Yard Murder.
1883		Keighley Midland Railway Station opened.
		Great Northern Railway Keighley to Queensbury opened.
1887	Xmas	Keighley Market lighted with Electricity.
1888	September 4th	Devonshire Park opened.
1889		Engineers' strike August 19th to September 19th.
		Old Horse Trams started in Keighley – Discarded 1904.
1890	June 26th	Bingley Cottage opened.
1891	July 21st	Lund Park opened.
1892		Keighley Post Office opened and Victoria Park opened July.
	June 30th	Mechanics Clock started.
1894	December 20th	Anderton Chimney Blown Down Keighley.
1896	December 6th	Keighley Temperance Hall opened.
1897	February 3rd	Morton Banks Hospital opened.
1898	December 6th	Morton Banks Boiler explosion.
1904	April 14th	Firth Mill Fire £4200 damage.
	October 3rd	Charlie Elliot born. Gordon Richards born. Silsden Gas Works opened.
1906	June 11th	Haworth Lady Parachutist fell to her death.
		Harry Myers, Keighley Footballer injured at Dewsbury. Died December 19th – Buried on Sunday.

1907		Buffalo Bill visited Utley.
1908	April 11th	Wadsworths Printers Russell Street fire - 2 killed.
1909	June	Myrtle Park Bingley opened.
1912		National Health came in force June.
1916	June 5th	Lord Kitchener drowned on HMS *Hampshire*.
1922	May 22nd	Horatio Bottomley sentenced to Maidstone Jail. Released July 29th 1927.
1926	February 4th	Last Tram rail took up in Keighley by Alderman Arthur Smith.
1927	March 2nd	Keighley Butcher gassed in Kings Arms Yard.
	July 22nd	Woolworths opened in Keighley.
	July 27th	Freddy Welsh died in USA.
1928	March 21st	New Bridge at Stockbridge started. Opened to traffic September 4th 1930.
1929	November 25th	Angel Inn closed (Falling Down)
	June 29th	Riddlesden Murderer sentenced during His Majesty's Pleasure.
1930	March 13th	Guide Inn opened.
	October 13th	Pew Abbott opened Hardings Lane.
	July 28th	Oliver Preston Murdered, Gill sentenced to death December 30th.
1934	August 4th	Fleece Hotel Low Street closed.
	December 17th	Robots opened top of Cavendish Street and the Cross.
	December	Market pumps taken out.
1935	March 29th	Marks & Spencers opened in Keighley. Bradford opened March 30th

Societies at that time rarely employed their own staff surveyor and we received a large volume of their valuations which we tried and in the main, succeeded in giving a forty-eight hour service. We also did a fair amount of rating of properties and it was somewhat easier than it is today.

The position generally in 1938 was not easy. At first we had threat of war, then Munich when Neville Chamberlain returned, declaring that it was to be peace in our time. A very good thing before the war was de-rating. If you were a manufacturer you only paid ¼ rates; the scheme was devised to help those who actually manufactured rather than those who imported.

Cover of the Property Guide 1938

Some of the entries reproduced here and on the following pages make interesting comparisons with today's prices

ILKLEY and BEN RHYDDING
Houses for Sale

A centrally situated Stone-built House, containing:—On the Ground Floor: Vestibule, Entrance Hall, Kitchen, 2 large Reception Rooms. On the First Floor: 3 Bedrooms, Bathroom and W.C. On the Second Floor: 4 Bedrooms and Boxroom. In the Basement: Kitchen, Coal, Keeping and Wine Cellars.
Ref. No. I/21/12.   Rateable Value £30.   Price **£450.**

Centrally situated Stone-built Terrace House, in a good residential district, comprising Entrance Hall, 2 Reception Rooms, Kitchen, Scullery, Coal-place, etc., 4 Bedrooms, Bathroom and W.C. Electric light. Small Garden.
Ref. No. I/14/1.   Rateable Value £21.   Price **£500.**

A Stone-built End Terrace House, situated in a very convenient position. The accommodation comprises Entrance Hall, Dining Room, Drawing Room, Kitchen, Scullery, Coal-place, 4 Bedrooms, Bathroom and W.C. Electric light is installed. and there is a Garden at both front and rear.
Ref. No. I/2/4.   Rateable Value £28.   Price **£500.**

An excellent Semi-Detached Residence, built of Stone with slate roof, and in a very convenient situation, being only 5 minutes' walk from station and shops. The accommodation comprises Entrance Hall, 2 Reception Rooms, Basement Kitchen, Scullery, Larder and Coal-place, 4 Bedrooms, Bathroom, 2 W.C.'s. Electric light. Small Garden at front and rear.
Ref. No. I/5/2.   Rateable Value £27.   Price **£500.**

SKIPTON—Houses for Sale

An excellent Stone-built Terrace House, containing:—On the ground floor: Entrance Hall, Dining Room, Drawing Room, Kitchen. On the first floor: 3 Bedrooms, Bathroom. There is also a good Attic and Wash Cellar. Electric light is installed throughout.
Ref. No. S/21/8.　　　　　Rateable Value £18.　　　　Price **£650**.

A well-built Stone and slated Terrace House. The accommodation comprises Entrance Hall, Drawing Room, Dining Room, Kitchen, 2 Bedrooms, Attic Bedroom, Bath-room and W.C., Cellar and Store. There is also a small Garden. Electric light is installed.
Ref. No. S/1/11.　　　　　Rateable Value £15.　　　　Price **£650**.
　　　　　　　　　　　　　　　　　Or To Let at **£1** p.w. & rates.

An excellent Stone-built Terrace House, containing—On the Ground Floor: Living Room, Sitting Room, Kitchen with Yorkist range and h. & c.; on the First Floor: 3 Bedrooms. There is also a good Attic, Coal and Keeping Cellars and Store. Electric light is installed, also there is an outside W.C.
Ref. No. S/10/3.　　　　　Rateable Value £12.　　　　Price **£650**.

A Stone-built and slated End Terrace House, containing Entrance Hall, Dining Room, Drawing Room, Kitchen, 3 good Bedrooms, Bathroom and W.C. There is a Greenhouse, a Boiler-House, a Garage, a small Workshop and a Coal-place.
Ref. No. S/16/4.　　　　　Rateable Value £24.　　　　Price **£750**.

An excellent Stone-built and slated Boarding House, near the centre of Skipton. The accommodation comprises Entrance Hall, Dining Room, Sitting Room, Scullery, Pantry, Store, 4 Bedrooms, Combined Bathroom and W.C.
　　　　　　　　　　　　　　Rateable Value £22.　　　　Price **£775**
Ref. No. S/19/4.　　　　　　　　　　　　　　　　and **£150** for business.

A new part Stone and rough-cast Semi-Detached House, in Skipton's new residential district, with a large Garden at the front. The accommodation comprises Lounge with tiled hearth, Oak Mantel and fireback boiler, Dining Room with bay, tiled fireplace and raised hearth, half-tiled Kitchenette, Pantry with tiled slab, three Bedrooms, two with fires, Bathroom, separate W.C., two Stores. Built-in Garage. Gas and electricity are installed. The house will be entirely decorated to suit the purchaser, and there are no liabilities for roads, etc.
Ref. No. S/24/3a.　　　　　　　　　　　　　　　　　　　　Price **£875**.

A House exactly similar to the above, fitted out as a show house.
Ref. No. S/24/3b.　　　　　　　　　　　　　　　　　　　　Price **£875**.

HOTEL TRANSFER

28

An excellent Stone-built Detached Residence, situated in a very pleasant position in Pool-in-Wharfedale. The accommodation comprises Entrance Hall, Dining Room, Drawing Room, Scullery, Kitchen and Larder, W.C., 5 Bedrooms, Bathroom, separate W.C., Attic. Electric light is installed. To front and rear is a good Garden of nearly 1 acre, and comprising Tennis Court, Kitchen Garden and Orchard, with Wooden Garage, Summer-house and Tool-house. Rates approx 12/- in the £.
Ref. No. O/7/8.　　　　　Rateable Value £52.　　　　Price **£1,750**.

A Brick-built and rough-cast Semi-Detached Residence, situate in Pool, and containing Dining Room, Drawing Room, Kitchen, Scullery, three Bedrooms, Bathroom and separate W.C., two Bedrooms, Store Room. There is a beautiful enclosed Garden with room for garage. Electric light is installed.
Ref. No. O/5/4.　　　　　Rateable Value £34.　　　　Offers required.

KEIGHLEY—Houses for Sale

A well-built Scullery House, containing Reception Room, Cellar, Wash Kitchen, Bedroom and Attic Bedroom. Gas is installed.
Ref. No. K/1/10.   Price £250.

A Stone-built and slated Terrace House, containing Tiled Entrance Passage, Drawing Room, Living Room, Scullery, Wash-Cellar, Keeping and Coal Cellar, 2 Bedrooms and a store. There is also an Attic-Bedroom.
Ref. No. K/1/6.   Rateable Value £15.   Price £280.

A Stone-built Terrace House, containing Entrance Passage, Drawing Room with bay window, Sitting Room and small Scullery, two Bedrooms, Combined Bathroom and W.C., Attic Bedroom, Cellar Kitchen with range, Wash Kitchen and Keeping Cellar. Electricity is installed.
Ref. No. K/2/4.   Rateable Value £19.   Price £425.

A Stone-built Terrace House, overlooking Victoria Park, and containing Entrance Passage, Sitting Room with bay window, Dining Room, Kitchenette, two Bedrooms, Combined Bathroom and W.C., 2 Attic Bedrooms, Cellar Kitchen, Wash Kitchen, Scullery and Outside W.C. Electric light.
Ref. No. K/2/3.   Rateable Value £22.   Price £500.

Stone-built Terrace House, containing Entrance Passage, Drawing Room, Dining Room, Kitchen, Cellar Kitchen, Wash Cellar, Keeping Cellar, Coal Cellar, 3 Bedrooms, 2 Attic Bedrooms, Bathroom and W.C. Gas is installed.
Ref. No. K/10/6.   Would consider letting.   Price £700.

A Stone-built and slated Dwelling-House, accommodation as follows:— Entrance, Drawing Room, Sitting Room and Kitchen. On the first floor: 3 Bedrooms, Bathroom and W.C. Also good dry Cellars. Electricity is installed.
Ref. No. K/19/2.   Rateable Value £21.   Price £700.

OTHER DISTRICTS—Houses for Sale

BRADFORD.
A substantial Stone and slated Terrace House, containing Entrance Hall, Sitting Room, Living Room, Cellar Kitchen, Coal and Keeping Cellars, etc., 2 Bedrooms, Bathroom, 2 Attic Bedrooms. Electric light and gas are installed. There is a small Garden to front and rear, and an outside W.C.
Ref. No. O/16/6a.   Offers required.

GRASSINGTON.
A charming Stone-built Cottage, situated between Grassington Moor and Coniston Moor, with extensive views to the South-East towards Greenhow Hill. This would form an ideal Summer or Week-end Cottage. The accommodation comprises 2 Reception Rooms, 5 Bedrooms, Kitchen and Scullery. Good Garden.
Ref. No. S36/17/2.   Rateable Value £7.   Price £250.

BINGLEY.
A well-built End Terrace House, containing Living Room, Kitchen, 2 Bedrooms, 1 Attic, W.C., Coal and Keeping Cellars.
Ref. No. S/22/2a.   Rateable Value £8.   Price £250.

PROPERTY SALES BY AUCTION UNDERTAKEN

Investment Property and Business Premises For Sale

HAWORTH.
A Shop with house attached, 15ft. frontage to Main Street, is Stone-built, with flagged roof. Living accommodation as follows: Living Room, Cellar-Kitchen, 2 Bedrooms, lean-to Wash-house, Outside W.C. and Coal-place. Useful Garden to the rear. Gas and electricity installed.
Ref. No. K/3/1.   Rateable Value £12.   Price **£350.**

OTLEY.
Substantial Stone-built Terrace House, very conveniently situated in Otley. The accommodation comprises Living Room with cooking range, Scullery with sink and set-pot, 3 Bedrooms, W.C. Coal and Keeping Cellars. This is excellent investment property, being in good condition and at present let at **8/4** per week and rates.
Ref. No. O/4/14.   Rateable Value £10.   Price **£350.**

OTLEY.
Valuable Freehold Investment Property, comprising Two Stone-built and slated Terrace Houses, in excellent condition, and each containing Entrance Passage, Sitting Room, Living Room with range, Kitchen with sink and h. & c., 2 large Bedrooms, Bathroom, 2 Attic Bedrooms, outside W.C. to rear, and useful Cellarage. There is a small Garden to front and rear of these properties. The houses are let to good tenants at a rent of 10/- per week, the tenants paying their own rates.
Ref. No. O/23/3.   Rateable Value £16.   Price **£350** each.

MORLEY.
Two Brick-built and blue-slated Residences, each containing Entrance Hall, Sitting Room, Living Room, Kitchenette, Larder, Coal-place, Landing, 3 Bedrooms, Combined Bathroom and W.C. There is a Garden to front and back, with ample room for the erection of a garage. Electric light is installed and gas laid on. Each house is let on a weekly tenancy, producing a nett annual rental of **£30 0s. 2d.** each.
Ref. No. O/3/26.   Price **£400** each.

MORLEY.
A Brick-built and rough-cast blue-slated Residence, containing the following: Passage, 2 Sitting Rooms, Kitchenette, Larder, Coal-place, Landing, 3 Bedrooms, Combined Bathroom and W.C. There is a Garden to front and rear, with room for the erection of a garage. Electric light is installed and gas laid on. The house is let on a three years' agreement, producing nett annual rental of **£32 19s. 7d.**
Ref. No. O/3/26.   Price **£425.**

OTLEY.
A Stone-built and slated House and Shop, right in the centre of Otley, containing a Sales Shop, Living Room (with cooking range, sink, h. and c. water), Cafe, Bedroom with bath, 2 Attic Bedrooms, and a Basement. Electric light installed. Vacant possession on completion.
Ref. No. O/16/7.   Rateable Value £12.   Price **£450.**

OTLEY.
2 Stone-built Through-Houses, with slated roofs, near the centre of Otley, each containing a Sitting Room, Living Room with range and sink, 2 Bedrooms. There is one Outside W.C. for the two houses.
Ref. No. O/5/3.   Rateable Value £7 each.   Price **£450** each.

GOOLE.
A House and Shop, situate in Goole, and containing Kitchen, Dairy, Sitting Room, Sales Shop for Fish and Chips, 3 Bedrooms. Gas and electric light. This property is let at a weekly rent of 15/- net.
Ref. No. O/24/3.   Price **£450.**

RATING SURVEYS AND APPEALS

TO LET

SKIPTON.
Compact Weaving Shed 101ft. x 43ft. 6in., capable of holding a minimum of 48 looms of 40in. read space, up to 60 looms. Rent includes Power, Heat, Lights and Rates. Rent of Looms 6d. per week per loom.
Ref. No. S/8/5b.   Rent **1/6** per loom space per week.

OTLEY.
Within 2 minutes' walk of the centre of Otley, Valuable Stone-built Warehouse To Let.—For further particulars as regards rent, apply to the Otley Office.
Ref. No. O/9/2.

OTLEY.
Second Floor Office, with electric light installed. This Office is in the centre of Otley.
Ref. No. O/21/30.   Rent **5/-** p.w. clear.

SKIPTON.
A desirable Office To Let (10ft. 6in. x 9ft. 6in.). For further particulars apply Skipton Office.
Ref. No. S/17/4.   Rent **£17** per annum.

A Sales Shop and House, containing Sales Shop, Sitting Room, Kitchen with sink, two Bedrooms, Bathroom, Store Shed and Coal-place, W.C.
Ref. No. I/11/29.   Rent **8/-** p.w. & rates.

KEIGHLEY.
Ground Floor Workroom, 35ft. x 17ft. (approx.). Would make good Garage. Central position.
Ref. No. S/19/3a.   Rent **£22** per annum.

SKIPTON.
An excellent Stone-built Workshop, consisting of 2 ground floor rooms and 2 above.
Ref. No. S/3/3b.   Rent **10/-** p.w. clear.

OTLEY.
Excellently situated Lock-up Shop, in Otley. The accommodation comprises Sales Shop with large display window, Living Room or Store-room; there are also 2 Store-rooms on the First Floor, and there is an outside W.C. to the rear. Electric light is installed, and gas is laid on.
Ref. No. O/2/7.   Rent **£24** p.a. & rates.

MENSTON.
A Stone-built House and Shop, centrally situated in Menston, and containing Sales Shop with good display window, Window Fixtures and Fittings, good Counter, large Bedroom or Living Room, large Bedroom and W.C., Cellar with sink and Coal-place. Electric light is installed.
Ref. No. O/15/4.   Rateable Value £19. Rent **£26** p.a. & rates.

OTLEY.
A Terrace House and Shop, situate in one of the main streets of the town, and containing Sales Shop with display window, Pantry, Kitchen, Scullery, 2 Bedrooms, 2 Attic Bedrooms, Store Cellar, Outside W.C. There is both electric light and gas installed.
Ref. No. O/12/1.   Rateable Value £22. Rent **14/-** p.w. & rates.

ILKLEY.
A Corner Shop, facing Head Post Office, double-fronted and basement; electricity, water laid on, To Let on annual tenancy or lease.
Ref. No. I/20/1a.   Rateable Value £25.   Rent **£40 p.a.**

VALUATIONS FOR ALL PURPOSES

Farms and Land

OAKWORTH, near KEIGHLEY.
A useful Farm, with 15 acres of land. The house comprises Sitting Room, Living Room, Kitchen, 2 Bedrooms and Closet. The out-buildings include Store, Shippen for 6, Pigstye for 4, 2-stall Stable and Cart Shed. The Farm is let at £36 p.a. and rates.
Ref. No. S/10/11.   Offers required.

LONG LEE, near KEIGHLEY.
18 acre Farm, including 3¾ acres of excellent Building Land, ready for immediate development. The Farmhouse contains Sitting Room, Living Room, three Bedrooms, Boxroom. Also an Earth Closet.
Ref. No. K/18/1.   Price: Offers.

OTLEY.
5.445 acres of Land covered with bracken and rough timber.
Ref. No. OP/20/1.   Price £50.

SASKATCHEWAN, CANADA.
Valuable Mixed Farm of 320 acres, comprising 220 acres of Arable Land, on which wheat and oats are grown, and 100 acres of Wild Grass, with accommodation for 15 Cows for a Milk and Cream Round
Ref. No. O/14/16.   Price £600.

OAKWORTH.
18 acre Farm, with shippen for 10, Barn, Stable, three Pig Cotes and Garage. The accommodation comprises Living Room and Kitchen, two Bedrooms; electric light is installed.
Ref. No. K/22/1.   Rateable Value £6.   Price £750.

WILSDEN.
25 acre Farm, with Shippen for 12, Barn, Wash-house, 2 Pig Cotes, etc. The Farmhouse contains the following:—Sitting Room, Living Room, three Bedrooms. Also an Earth Closet.
Ref. No. K/18/1.   Price £1,200.

FOULRIDGE.
A very good 37 acre Farm 27 acres of which are pasture, and 10 acres meadow. The farm is now Let at a rent of £75 per annum.
Ref. No. S36/19/1d.   Price £1,400.

FURNITURE COLLECTED FOR THE SALEROOM

BUILDING LAND

OTLEY.
1,084 sq. yds. of Valuable Building Land in the centre of Otley.
Ref. No. O/17/14.   Price £3,500.

BINGLEY.
18.13 acres of valuable Building Land, facing main road, the whole area to be sold, and not in plots.
Ref. No. K/5/1.   Price 1/- sq. yd.

BURLEY.
About 4 acres of good Building Land, very well situated, in Burley.
Ref. No. O/18/5.   Price 1/3 per sq. yd.

BURLEY-MENSTON ROAD.
2,386 square yards Building Land.
Ref. No. O/4/12.   Price 1/6 per sq. yd.

MENSTON.
7,845 square yards of Valuable Building Land in Menston.
Ref. No. O/5/10.   Price 2/- per sq. yd.

BRAMHOPE.
70 Acres of Valuable Building Land, excellently situated, and suitable for houses of the £550 to £750 type.
Ref. No. O/22/19.   Price 2/- per sq. yd.

MENSTON.
20 acres of excellent Building Land, with splendid views over Burley and Menston district. Can be sold in plots.
Ref. No. O/16/10.   Price 2/- per sq. yd.

POOL.
2 acres of Valuable Building Land, in Pool.
Ref. No. O/22/1.   Price 3/- per sq. yd.

WEETON.
1,000 square yards of Valuable Building Land at Weeton. Well situated.
Ref. No. O/3/13.   3/6 per sq. yd.

ILKLEY.
Valuable Freehold Building Land in an elevated position in Ilkley.
Ref. No. I/23/9.   5/- per sq. yd.

HOTEL TRANSFER

The Property Guide advertised on page one the Yorkshire Insurance Co which had been started in 1820, the same year that Dacre & Son had been founded. The advert stated you could purchase your house with a loan of up to 75% of its value at 4½% interest. The advert was clear and concise, unlike some of the terms offered by present day financial experts.

The staff at the Keighley office including myself, comprised Bill Spencer the cashier who had originally joined the firm in 1936 when he left the wool trade at £1 per week for a more secure job with Dacres at 10 shillings a week. He stayed with the firm for fifty years. The staff was completed by a young girl who acted both as receptionist and typist, including filing and making tea. Looking back it was quite remarkable how the office grew as fast as it did.

HOUSE PURCHASE
Without Anxiety

BUY YOUR HOME with the help of The

Yorkshire Insurance Company Limited

•

LOANS GRANTED UP TO
75% AT 4½% INTEREST

Under the Company's attractive House Purchase Scheme, the risks that dependants may be burdened with a debt on the house is avoided . . . .

Write for Particulars, using the enclosed Business Reply Card, or call at the Company's Branch at
INFIRMARY STREET :: LEEDS

It was before the days of electric typewriters and recorders and I was only twenty-three years of age, but arriving at the office at 8 am and working a minimum of ten hours on a normal day ensured a steady growth of the business.

One elderly client whom I well remember was Frank Green of York who had moved to Somerset. I was then in my early twenties and had been recommended to him. He entrusted me with the management of several estates in Leicestershire and Somerset. My fee was 10% of the rents received, £5 per day when I was doing his work and 3rd class railway fare when I travelled for him.

Frank Green

CHAPTER TWO

Wartime

Dacres had been in Keighley only eighteen months when war was declared. We expected a fall off in work but it was not so and we had more professional instructions than before.

The rateable value of a property is based on the rental value and Public Houses are no different. Just before the war the Robinson Case altered the position. Most breweries created tied houses letting tenants into the pubs and charging a fair rent, but then tying their tenants to buy all drinks, even mineral waters, from the brewery and fixing their own prices. In other words the breweries were adding to the rent by this profit on the tie. A free house could buy where they wanted at a lower figure. The Robinson Case decided that in arriving at the rental value for rating, the assessment had to take into account the increased rent a brewery would pay, as against the individual tenant.

At that time the County Valuer for the West Riding of Yorkshire was a surveyor called Lawrence Tattersall. The County Council levied a rate but it was included in the local authority demand and the local authority collected it and paid the money over to the County. Tattersall immediately served a proposal for an increase in the rating assessment of every public house in the West Riding, giving his reason that the assessment should increase to comply with the Robinson Case. For some reason the Keighley Appeals were heard before most of the others in the County and I was appearing for a local brewery and several owner occupied public houses. The War had just started and breweries were having difficulties supplying pubs and were not looking for fresh outlets. I pushed the point to the Local Assessment Committee who were at the time dealing with the matter, saying that the appeals were so worded to restrict any increase to the Robinson Case and due to the War did not apply. The Assessment Committee accepted my submission and Mr Tattersall thought it wise to withdraw all his hundreds of appeals.

Contrary to expectations Dacre, Son & Hartley found that the War produced a large increase in work which, coupled with staff leaving for military service, was not an easy situation. My day at the office began at 8 am and often went on until 9 or 10 o'clock at night.

An early instruction came from Wilfred Fattorini of Carleton near Skipton who was very patriotic and at the outbreak of war, had informed the Government that he had some spare warehousing and offered it to them. It was accepted by the Ministry of Aircraft Production for Rover Motors but at the last minute an official called to say that Rover were taking over the whole building, which they duly did. This put Wilfred Fattorini out of business and he had to dismiss his three-hundred-and-fifty employees and sell his stock at knock-down prices.

A few months later the Ministry informed him that the wall of the mill dam had collapsed and demanded that he repair it. I wrote back to the Ministry and pointed out that all services were included in the Requisition Order and therefore it was the Ministry's problem. They replied that Rover did not need the dam and therefore it was not included in the Order. Fortunately I knew that Rover occasionally used the dam for fire drill and sent Rover a bill for water. They fell for the bait and replied that they were not going to pay as the use of the dam was included in the requisition. The Ministry was cornered and eventually paid up, but what a terrible waste of time.

We soon became heavily involved in work connected with the requisition of property. The first instructions were from the Ministry of Food in connection with a slaughterhouse at Settle, North Yorkshire. On arriving at the premises I found the building little more than a large barn with an iron ring in the floor for tethering beasts, a block, a humane killer and a few knives. I was finished and back home in less than two hours. The Ministry's own scale of charges for this work was £120 based on cubic contents, not bad for two hours work in 1939 and when someone from the Ministry telephoned me to raise the matter I agreed that the fee was excessive but it was based on their own scale. We agreed an amended sum of £60 and I sent in our account. After a short while we received another telephone call from the Ministry saying they still thought that account rather high. I pointed out that £120 was the fee according to their scale but I was willing to accept any sum they thought suitable. Finally a fee of £30 was agreed and this was paid promptly. The only trouble was that £30 arrived for the next three months and trying to stop them from sending it was harder than trying to stop Reader's Digest. In the end the cheques stopped arriving as mysteriously as they had started but I had received the original fee of £120 in instalments.

Another interesting case concerned a large disused textile mill which the Ministry of Food had requisitioned for storage. In this instance I acted for the owner and when he handed me the requisition papers I saw that the Ministry had described the mill as premises, together with the shafting and pulleys already present and fixed on the premises.

The question of shafting did concern me and I told my client that requisitioning of the mill was straightforward, all that was needed was to prepare a Schedule of Condition to be agreed with the Ministry and negotiate the amount of rent, but shafting, pulleys, belting etc were not part of the freehold, but chattels. These were different and could not be requisitioned but had to be bought. I was very concerned and made the position quite clear to my client. There was a war on and I suggested we should tell the Ministry of Food our thoughts on the position and if they agreed with our view we would cooperate with them in putting the matter right. The Ministry of Food agreed with us, derequisitioned the premises and then requisitioned the mill alone excluding the shafting etc. I think my client was well satisfied for when the War ended and the mill was derequisitioned I dealt with the dilapidation claim and when it was all finished and I had been paid, the client called at my office to give me a gold cigarette case and £25 for my articled clerk.

The War Department Land Branch sent us a list of requisitioned properties at frequent intervals and our job was to prepare a Schedule of Condition of the buildings and organise the owner or his agent to sign it. The object was to help if a dilapidation claim was made on derequisition.

It soon became clear that I had more work than I could do so I resigned my agency work in Lincolnshire, Leicestershire and Somerset. A little later I thought the Ashwich Estate in Somerset was past history and it was with some surprise that I received an urgent message one Friday from the owner Frank Green asking me to call and see him. He would not give a reason for the meeting, only that it was secret and involved the military. The following day I caught a train from Leeds in the late afternoon to Bristol and settled down to a tedious journey to the West Country, which with luck would reach my destination before midnight, without the convenience of food either on the station or on the train. It being a Saturday night there were not many passengers travelling and the journey was uneventful until approaching Birmingham when suddenly the train came to a halt. Our arrival had coincided with the first 1,000 bomber raid on that City and a night I shall never forget.

The guard came along the train advising passengers for Birmingham to leave the train and find any shelter they could. The signal box on Birmingham station had been hit, causing obvious complications but as soon as things had quietened down he assured us they would try and reverse the train and find an alternative way round the City and continue on to Bristol. I managed to reach Frank Green for lunch on Sunday and after the meal the reason for my visit was revealed. It seemed that the War Department wanted to site some anti-aircraft guns on Exmoor on land forming part of his estate. He was quite willing to let them and in truth he could not have stopped them even if he had wanted to. Furthermore he had no wish to charge them rent for the land. What was worrying him was the possible difficulty of getting his land back again after the War and he wanted me to look after his interests.

In fact the whole journey had been a bit of a waste of time, certainly not worth sitting on a train underneath 1,000 bombers dropping their cargo. The matter could have easily been sorted out by correspondence but at least Frank felt better. I said goodbye to him before he retired to his room for his afternoon sleep and an estate worker drove me back to Taunton to catch a train to Bristol which would connect with the Leeds train.

A firm in Croydon called NSF, Nuremberg Switch Factory, was a German firm working on vital electronics and radar technicalities. The management was German and naturally the weekend before war was declared, returned home leaving no-one really in charge. The Ministry of Aircraft Production took action and a firm called Smart & Brown Engineers agreed to take over the management of NSF. When bombing started the Germans, knowing the location of the factory and its contribution to the war effort, made it the target of the first daylight bombing raid on London. The premises were completely destroyed and thirty-five of the three hundred employees were killed.

I had just finished lunch one Saturday when a Mr Dean of the Ministry of Aircraft Production telephoned and explained the NSF problems. He told me that the Ministry had requisitioned Trico at Keighley and part of Dalton Mills for NSF. The firm's managers were even then travelling north and were anxious to arrange a meeting with me on the Sunday morning to discuss details of the move.

The Chairman of Smart & Brown was an elderly gentleman called Wilding Cole and he was excellent to deal with. It appeared that the plant salvaged from the wreckage at Croydon was already on its way and the first load could be expected within 36 hours.

Things move fast in wartime. Fortunately both Trico and Dalton Mills were already clients of ours and Schedules of Conditions could be agreed almost immediately. It was not just a matter of plant and machinery: two coachloads of key workers and their families were expected to arrive the following evening in need of housing. Keighley Corporation were contacted and offered help by requisitioning housing.

On the following weekend it was the only day in my life that I opened our office on a Sunday, but permanent housing had to be found for the NSF staff. We also agreed to make arrangements for their furniture to be brought up from London. This is normally outside the bounds of a surveyor's duties but under the circumstances we felt that we must tackle the problem. We did not realise how difficult it would be to obtain removal vans. Every big firm we rang refused to go into London and those already there were too busy, but eventually a local Keighley man with three vans agreed to do the work at £25 per trip. Those were the days.

We had carried out work for the Army, Ministry of Food and Ministry of Aircraft Production but it was a surprise when we received a call from the Admiralty enquiring if we would act for them in the requisitioning of a large mill at Cornholme near Todmorden. I agreed I would do the job for them but was very short of petrol, whereupon I was asked how many gallons I would need to complete the work. I estimated twenty-five gallons and was told it would be arranged. The following day coupons arrived for the said twenty-five gallons, furthermore coupons to the same value arrived monthly to do the job. Somebody had thought the request from the Admiralty to the Petrol Office was for a monthly allowance and not a one-off matter and I was not going to complain.

I was surprised to learn that the mill was to be used for the storage of ships' masts, the reason for the decision being that Todmorden was at the centre of England and therefore equidistant from all ports. Cornholme Mill was the only job I did for the

Admiralty but I think they must have been satisfied for shortly after completing the matter, I received a further telephone call from them enquiring whether I would be prepared to go as a civilian to Aden, dealing with property matters on their behalf, Aden being of course a neutral country in the Gulf. They even discussed salary but I told them I could not carry out the work for two reasons: firstly I had just got married and secondly I had a job keeping Dacre, Son & Hartley going in order that there would be work for those members of staff returning to the firm at the end of the War.

As well as all the work that we carried out for the various services in connection with the War we had our own general practice to run, including valuations for building societies and others and probates and sales of all types of properties.

Finally came the controversial dropping of the first ever atomic bomb on Hiroshima, Japan on the 6th August 1945, closely followed by the second one that devastated Nagasaki. The War was over and we began preparing for peace.

CHAPTER THREE

Post-war

The end of the War saw a big change for Dacre, Son & Hartley. During hostilities the firm had been run by my brother Douglas aged thirty-four and myself aged thirty, together with three offices at Otley, Ilkley and Keighley assisted in the main by a temporary staff of school-leavers who were with the firm for a short time before joining the services. My brother and I had each been putting in at least ten hours work a day including weekends and taking no holidays.

The first asset when peace came was the return of our partner Douglas Hartley who had been engaged in Hull dealing with war damage compensation on behalf of the Government. It meant that we now had a partner at each of our three offices and it also meant that the partners could return to the steady growth of the firm that had been taking place prior to the War. It gave great satisfaction to the partners to be able to give their jobs back to all the returning staff who had been on active service.

Owing to the increase in the value of the housing market auction sales were favoured by many and we were holding these sales almost weekly.

Each partner ran his own office but there were certain rules that applied to all, such as office hours, scale of charges, all letters to be answered the same day and signed by a partner, all work to be carried out free of charge for charities unless there was money coming in, for such as property sales.

At Keighley whenever an articled clerk came to the firm on his

**THE VICTORIA HALL SALE ROOM,
ILKLEY.**

**FORTHCOMING SALES
BY
DACRE, SON & HARTLEY.**

April 10th. — Central Business Premises, House and Shop, No. 45, High Street, Yeadon (with Vacant Possession). Solicitors: J. R. Philips and Co., Bradford.
April 11th. — 4, Wardman Street, Keighley. Solicitor: J. Rhodes Clough, Keighley.
April 11.—" Sylva Dene," 2, Poplar Drive, Sandbeds (with Vacant Possession). Solicitors: Wright and Atkinson, Keighley.
April 11th. — 116, South Street (Vacant Possession). Solicitors: Wm. Dewhirst and Sons, Keighley.
April 11th.—40, Catherine Street; 217, South Street; 8, Hope Place; and 4, The Walk, Keighley. Solicitors: Turner and Wall, Keighley.
April 11th.—83, Devonshire Street West, Keighley. (Vacant Possession.) Solicitors: Burr, Sugden and Co., Keighley.
April 11th.—57, River Crest, Bradford Road, Riddlesden, Keighley (Vacant Possession). Solicitor: J. Rhodes Clough, Keighley.
April 11th.—27, Prospect Place, South Street, Keighley. Solicitor: J. Rhodes Clough, Keighley.
April 11th.—8, Victoria Road, and 4, Arcadia Street, Keighley. Solicitors: Wm. Dewhirst and Sons, Keighley.
April 12th.—Four Cottages, Rose Terrace and Victoria Avenue, Addingham. Solicitors: Waddington, Clapham & Feather, Keighley.
April 12th.—15, Wharfedale Drive, Ilkley (with Vacant Possession). Solicitors: Atkinson, Dacre, and Slack, Otley.
April 13th. — Detached Residence, Shop, Land and Premises, Cambridge House, Otley. Solicitors: Atkinson, Dacre and Slack, Otley.
April 13th. — Detached Residence, Willowbank, Leeds Road, Otley (with Vacant Possession). Solicitors: Atkinson, Dacre and Slack, Otley.

April 13th.—Terrace House, No. 17, Ashfield Place, Otley. Solicitors: Atkinson, Dacre and Slack, Otley.
April 13th.—95, Albion Street, Otley (with Vacant Possession). Solicitors: Barret, Chamberlain & McDonnell.
April 13th.—19, Albion Street, Otley (with Vacant Possession). Solicitors: Atkinson, Dacre and Slack, Otley.
April 14th.—31, South View Terrace, Silsden (Vacant Possession), together with furnishings. Solicitors: Wm. Dewhirst and Sons, Keighley.
April 18th.—Valuable Furnishings and Effects at the Victoria Hall Saleroom, Ilkley.
April 20th.—Moana, Strathmore Road, Ben Rhydding (with Vacant Possession). Solicitors: Gaunt, Fosters, and Bottomley, Bradford.
April 21st.—10, Skipton Road, Steeton (with Vacant Possession), also remaining furnishings. Solicitors: Wm. Dewhirst and Sons, Keighley.
April 23rd.—12, Russell Street, Skipton (with Vacant Possession). Solicitors: Bannister, Bates and Son, Morecambe.
April 25th. — 9 and 10, Damems, Keighley (No. 10 Vacant Possession). Solicitors: Turner and Wall, Keighley.
April 25th.—Cragg Lea, Flappit Springs, Keighley (Vacant Possession). Solicitors: Turner and Wall, Keighley.
April 25.—1, 3, and 5, Marley Street, Keighley. Solicitors: Turner and Wall, Keighley.
April 25th.—17, Aireworth Street, Keighley. Solicitors: Wright and Atkinson, Keighley.
April 25th. — Central Business Premises, Keighley. Solicitors: Wm. Dewhirst and Sons, Keighley.
April 28th.—Poultry houses, equipment, etc., Brooklands Farm,

Properties being advertised in the Keighley News *1945*

first day at the office he was given a copy of the 'Ten Commandments':

TEN COMMANDMENTS

1 Organise mind and thoughts so that you can plan your day.

2 Aim to do the job in hand better than anybody else – then you may do it as well.

3 Always do what you think is right, *without regard to financial results.*

4 Never consider a job too small.

5 If a mistake is made, admit it – good people make mistakes once but they do not make the same mistake twice.

6 People who come to work late spend all day trying to catch up.

7 Never drink alcohol in the day-time when you are working.

8 Pay all your bills promptly.

9 If you do not know how to do a job – do not be afraid to ask.

10 If you are going to get into trouble – get into trouble for something you have done rather than something not done.

With the end of the War, town councils throughout the country made an attack on substandard housing. There were many large areas of Compulsory Purchase Orders dealing with what was referred to as Slum Clearance. These were back to back houses which lacked amenities, with stone flag floors, no hot water, no electricity and outside toilets, in many cases what were known as tippler toilets. A tippler was a toilet that had no independent chain but was flushed every time the kitchen sink was used. The sink waste when emptied, ran from the sink down a pipe to a hopper and when full the weight of the water worked the tippler which flushed the toilet. The toilets were either what were called long hoppers or short hoppers according to the distance the user sat above the hopper.

The first housing development was the appearance of prefabricated bungalows with a limited life followed by large Council estates, sadly all of a type that can easily be identified as a council house. With hindsight the estates would have been better with more variation of design and layout. To save on land use, multi-storey flats appeared and Keighley erected ten-storey flats on Park Lane on top of a hill, not the best of sites. All the new housing enabled the Council to speed up slum clearance. In a clearance scheme there were four types of properties:

1. Unfit houses.
2. Fit houses but needed for the Scheme.
3. Commercial properties.
4. Buildings used for a purpose that does not usually change hands eg churches and clubs.

Compensation varied according to its classification:

1. Unfit house – Site value plus twice rateable value if well maintained outside and further twice the rateable value if well maintained inside.
2. Fit houses – Full market value.
3. Commercial properties – Full market value.
4. Buildings used for a purpose that does not usually change hands – The cost of providing equivalent reinstatement. This is known as Rule 5.

Dacre, Son & Hartley at Keighley had a large management department and kept a sharp eye on which properties could be affected by slum clearance. By selling to the tenant of an unfit house for a figure of £400, the landlord giving a 100% mortgage, repayable at £1 a week, providing the deal had run for two years, the compensation would be increased from site value about £5 to market value, usually £400/500 at that time.

When a Compulsory Purchase Order was made for slum clearance it was policy to object, the Minister then dealing with the matter at a Public Inquiry presided over by a Minister's Inspector. This took time, giving landlords a longer period before loss of rent. It was usual at a Public Inquiry for owners to be represented by a solicitor who called a surveyor to refute the Council's claim to the property being unfit. This again delayed the matter until the Inspector's Report was received after many months.

One solicitor Jack Mewies, for which I gave evidence, was excellent with a good sense of humour. Once when going through the cross-examination of the Council's surveyor who was complaining that an unfit house had a broken stone flag garden path, Mewies keeping a straight face, asked the surveyor if he condemned all crazy paving. In the same Inquiry the Council surveyor had said that the unfit house we were concerned with had windows that did not open and there was a lack of air. The Inquiry was being held in Bingley Council Chambers and it was a very hot summer's day. I happened to see that the Council Chamber windows did not open so I passed a quick note to this effect to Mewies who immediately asked the Inspector if we could have a window opened as it was oppressively hot. The Inspector agreed and it was then realised that the windows would not open and Mewies asked the Council surveyor if the fact that the windows would not open made the Council Chamber unfit for human occupation.

The orders were always confirmed by the Minister subject to

any variation he made. We then had to meet the District Valuer of the Inland Revenue to agree compensation. I had an interesting case of a Club in a clearance area in Keighley. The Club instructed me to act and asked me if I would attend a Committee Meeting and explain their position to them. The meeting was on a winter's night and when I arrived I found Johnny Binns the MP for Keighley in attendance; he had been invited there to confirm the correctness of the information I gave to the Committee. Johnny Binns was a popular Labour man in the town and I knew him well. I told the Committee that their claim against Keighley Corporation should be based on Rule 5 Equivalent Reinstatement in an existing building if one was available, but if a suitable building was not available then they were entitled to have a new club built. The Committee Chairman asked Johnny Binns if he agreed with my advice which naturally he did. The meeting closed and Johnny Binns and I departed. As we left Johnny suggested we should go for a drink and I told him I knew just the place, so he jumped into my car and we drove two miles to Cullingworth. In we went for the drink and after a while Johnny made the remark that we were in a very good pub and asked what they called it. He was indeed shocked and wanted to leave at once when I told him we were in fact drinking in the Cullingworth Conservative Club.

I agreed with Mr Padget, the District Valuer that I would meet him at the Keighley Club in order that he could make an inspection prior to meeting later, hopefully to agree the compensation. Both Padget and I were somewhat taken aback when I introduced him to the Club secretary who immediately asked to be excused. A couple of minutes later he returned with a bottle of whisky, saying the Committee would be pleased if he would accept it as a gift. Naturally Padget refused the whisky, but the Club secretary persisted with the gift so Padget informed him that if the whisky was not put away at once he would walk out, report the matter to the Inland Revenue Superintendent Valuer and the next person the secretary could expect to see would be a policeman.

I found generally that Compulsory Purchase was dealt with fairly on all sides and negotiations were confidential. I dealt with many cases in the redevelopment of towns, one such case in Keighley was Percy Wilkinson an undertaker who had a furniture shop in the centre of Keighley. I put in a claim and in due course met the District Valuer and agreed compensation. A few days later I was talking to the Mace Bearer at the Town Hall and he told me he was surprised how well I had done for Percy Wilkinson. This shocked me and I asked him how he knew I was acting and how much Mr Wilkinson had received. Without a blush he told me he had read all the details on the Deputy Town Clerk's desk.

Percy Wilkinson once told me a true story of how the wife of one of his clients had died in Keighley and was to be buried in Morecambe. The hearse in which as undertaker Percy sat in the front, was followed by two mourners' cars behind. It was a hot day and having passed through Long Preston the hearse driver put his foot down and was able to reach the Public House at Austwick well ahead of the following mourners. The hearse driver and Percy were just coming out when the mourners saw them. They stopped and the deceased's husband invited the whole party in for a drink.

The following year the husband died and he left instructions to his executors that he wanted Percy Wilkinson to be the undertaker. He wished to be buried with his wife in Morecambe and requested the same arrangement regarding the Public House at Austwick. Regretfully on this occasion he would be left outside but he insisted his estate was to pay for the drink.

CHAPTER FOUR

Clearances

Shortly after the War Keighley Corporation made a Compulsory Purchase Order dealing with twenty houses at Thwaites Village; surprisingly a Methodist Chapel was included. The Trustees of the Chapel asked me to act for them and when they came to see me, told me that due to a nearby area being recently cleared, they were down to two members at the Chapel and the congregation on a Sunday was down to five or six. I enquired as to the future of the Chapel had there been no Compulsory Purchase Order and was not surprised when they said they would have had to close in any case and they were planning to do so.

I explained to the Trustees that they must keep the Chapel open as I proposed that a claim against the Corporation should be based on Rule 5. In order to receive the benefit of Rule 5, there were four conditions:

1 The property must be of a type that does not change hands in the open market, eg church or club.

2 The property must be used for the purpose claimed.

3 There must be an intention to move and not close. The cost of equivalent reinstatement is not necessarily building new premises, it can be the cost of converting an existing building if one is available, but if no such building is available then the amount is the cost of a new building.

4 At the Lands Tribunal Rule 5 is at the discretion of the Tribunal.

The more I thought about the matter I found it difficult to appreciate why the Council had included the Chapel in the

Scheme. It was obvious that quite apart from the Clearance Order the Chapel would close and if the Council had then wanted the site they could have acquired it at little cost.

The Trustees accepted my advice and I asked where previous members had moved to and where the handful of existing members were likely to go. The Exley Head Methodist Church some two miles away seemed to be the choice. I asked the Trustees to call a meeting of the members one Sunday after Service and I attended, suggesting to them that they have a Resolution they would like the Worth Village Chapel to be reinstated by a new Church built on land owned by the Methodists at Exley Head. Actually the site in mind was adjacent and owned by the Exley Head Methodist Chapel.

Having had the Resolution unanimously agreed by all five of them I filled in and returned the forms to the Council stating that I was claiming compensation under Rule 5 and the amount would be negotiated when I had full details from the architect. I then waited for the reaction from the District Valuer who would be negotiating on behalf of the Corporation. I had not long to wait. I was the only Chartered Surveyor in practice in Keighley and had hundreds of claims with Geoffrey Padget, the District Valuer at Keighley and used to meet him on most Fridays to deal with them. Worth Village Chapel received early treatment. Geoffrey Padget was a good Valuer, naturally I thought he was a bit tight with Government money and no doubt he thought I was a bit generous in my valuations. I used to reckon his offer should be doubled to arrive at a fair figure, and no doubt he thought my claim should be halved to arrive at a fair figure. The result in both cases was usually very close to the settlement.

The first question he was to ask was "How many members were there at the Chapel?" I told him very few but all church attendances varied over the years and a good Minister could fill it. Again who can tell, maybe after a while they could be filling the place having the Messiah there. The attendance had also suffered because of the neighbouring Clearance Order. The District Valuer said he wanted to inspect the site and we could talk again about my claim on my next visit. I looked forward to the visit waiting to see his reaction when he realised the proposed site was next door to an existing Chapel.

The following week at my meeting with the District Valuer who had made his inspection, he asked why there was a need for two Chapels next door to each other, surely the Worth Village members could join with Exley Head, they would probably be able to fit into one pew and with a smile, he added he was prepared to pay the cost of one pew. I had expected such an attitude and was prepared for it. I told Geoff Padget it was not as simple as he suggested, as Exley Head was a modern Methodist Chapel, whereas Worth Village was originally a Wesleyan & Primitive Chapel. I knew the Mr Padget and family were Roman Catholics and suggested to him that if the local Catholic Church was acquired under a Compulsory Purchase Order would he suggest that they move in with a Church of England, Methodist or Congregational Chapel? Worth Village Chapel wanted to be reinstated in their own Chapel where they could carry on with their own type of service. We both said we would consider the position and discuss the matter later.

I assumed that Geoff Padget would discuss the situation with the Superintendent Valuer and in the meantime I saw a local solicitor, John Lees who acted for the Methodists in Keighley. John Lees arranged for an appointment with Mr Frank Layfield, a leading QC in London, well known as an expert in compensation and planning matters, later to carry out the Layfield Report on Rating on behalf of the Government. Those meeting Mr Layfield were John Lees, Harry Baldwin, a dedicated life supporter of the Methodist Cause in Keighley and beyond, and myself. I was working in London at the time and on the day of the meeting, Baldwin and Lees travelled to London and came to my room. I was staying in Park Lane at the Playboy Club; let me hasten to add it was above the Club, the fee being £10 per night including bed and breakfast which was brought to my room not by a bunny girl, but a figure more like a pantomime dame. Lees, Baldwin and myself discussed our claim and then went to our meeting with Frank Layfield whom I found to be a perfect gentleman, understanding and most helpful and he seemed to consider our actions were correct.

We returned to Keighley and on my next visit to the District Valuer I told him we had been to see Frank Layfield who would be acting for us in the matter if we went to the Lands Tribunal. Whether it had any influence in the matter I did not know but we heard shortly from the District Valuer that Rule 5 Equivalent Reinstatement was accepted by the Council. The new Chapel plans were prepared by the architect Alan Eden, the design being such that when the dust had settled the new Chapel would make a perfect extension to Exley Head Chapel, which it did very soon.

Shortly after the Worth Village Case I joined the Methodist Church Property Division in Manchester and became much more involved in Methodist properties. The practice has always been for local Chapel Trustees to appoint their own surveyor but they can always ask the Property Division for help. There was a case where I was asked for help in Norfolk where the agent was a very good agricultural auctioneer, but he had little experience of Compulsory Purchase Orders. The Chapel stood in the path of a bypass and it was clearly a Rule 5 case, Equivalent Reinstatement. When I looked at the building I noticed within 300 yards there was an old Chapel and on enquiring I was told that it was also owned by the present Chapel Trustees who had let the building for storage purposes. My fear was that the District Valuer would agree Rule 5 but then go on to say a new Chapel was not needed, but that the old Chapel could be refurbished. The Trustees took my advice, sold the old Chapel and successfully claimed a new one.

Another interesting case was in Newcastle upon Tyne when a

Compulsory Purchase Order was made. A good local surveyor had rightly claimed Rule 5 and stated the amount of the claim would be negotiated. After a while Newcastle Local Authority realised that Rule 5 would be very costly and wanted to get out of a deal for the Chapel. They said that they were not proceeding with the purchase as the claim form stated that the amount was to be negotiated and so worded, it meant that there was no claim made within time. Consequently it was decided that the Council would withdraw. No blame was attached to the surveyor as in fact I had myself used similar wording when claiming for Rule 5. After the Newcastle case instead of the words "to be negotiated" I started quoting a figure of £1,000,000. It bore no resemblance to the settlement but it made the claim legal.

CHAPTER FIVE

Mosque

The old West Riding of Yorkshire has always attracted immigrants and refugees. Before and just after the First World War there were Russian, German and Polish people chiefly interested in textiles and they settled around Leeds adding much to the benefit of the City. Immediately prior to the Second World War there were Jews from Germany escaping from the treatment by Hitler and in the main they were good businessmen. Shortly after the War there were Ukrainians and then Pakistanis arriving and their housing requirements created a problem. They were an unknown risk and to obtain a building society mortgage was almost impossible, houses to let in the private sector were not available and adding to the difficulty was the problem of the type of housing required which was at the bottom end of the market, with a limited life.

I felt in Keighley that it was my duty to try to help and I suggested to suitable vendors trying to sell their houses, that they should offer a private mortgage. The average price for such a property at that time was £500 with no deposit payable, availing immediate occupation for the tenant and providing the owner with a 4% return in interest on the mortgage and repayment at the rate of £2 or £3 per week. I negotiated 1,400 such sales with the mortgagors making the repayments to Dacre's Keighley Office and no vendor made a loss. In today's climate I do not think such transactions would work so well. A few years ago a Ukrainian came up to me in Keighley town centre and personally thanked me for such a purchase. He had arrived at Dacre's office in Keighley and we had managed to negotiate a property for him; fifty years on he still remembered.

The big difference with the Pakistanis was their tendency to live within the same areas. In Bradford they gathered round Manningham while in Keighley they seemed to favour

Lawkholme Lane. They also retained their own style of dress which immediately set them apart, tended to have large families and in nature were slow to mix within the community but when they did integrate they added to the culture of the town.

The first mosque in Keighley was in Lawkholme, being two back-to-back houses knocked together and after a short while it was included in a Compulsory Purchase Order covering a clearance area. I was instructed to act for the Muslim owners of the mosque and I explained to the church managers the basis of Rule 5 and told them if it passed all the requirements then the compensation could be Rule 5, but if challenged it would be finally at the discretion of the Lands Tribunal. I put in a claim to the Council stating that I wished to claim under Rule 5 and the amount would be supplied later when the figure was available from the architect. Having regard to the North Tyne Case in Newcastle the claim would now be considered invalid because I had not stated a figure of money. It would have enabled the Council to withdraw the CPO if they thought it was going to be too costly. Having put in the claim I awaited hearing from the District Valuer and I had not long to wait. He did not like the idea of the Corporation having to bear the cost of building a new mosque to replace the old converted houses and I naturally pointed out the need to be mindful of not being accused of race discrimination. What had been good enough for the Methodists in Worth Village should be the same for the Muslims in Lawkholme. The District Valuer suggested that I should clear the matter with the Town Clerk and accordingly I wrote the following letter:

"Dear Sir,

As you are probably aware we are acting for the Muslim faith in negotiating compensation for the mosque which is being compulsorily acquired by the Corporation.

We take the view that the Compensation shall be based on

Rule 5 and the District Valuer we believe has accepted this basis with certain suggestions.

Our clients' present premises are a residence which they purchased at considerable expense to themselves, and adapted it to their purpose. The District Valuer has suggested that it might be proper for them to acquire another residence and similarly convert it to their own use. Of course we see nothing wrong with this, providing they can acquire another property and provided they can obtain planning permission to convert it. At the end of the day they are no worse off either in accommodation or financially than now. If our clients are fortunate enough to find another house that will readily convert and planning permission is forthcoming, we are then faced with the problem that mosques must face in a certain direction towards Mecca. Also we are faced with the problem that the congregation which numbers 600 for normal worship, have to pray five times a day, accordingly the mosque must be in a convenient place for our clients' congregation. We are then faced with the problem of parking, as obviously any church with a congregation of 600 does need parking facilities.

We shall be glad if you can let us know if you think the Corporation has any property which is suitable, or they can make any suggestions as to where we can find a house where we can get planning permission and which will convert at a reasonable cost, where it will point towards Mecca and at the same time is convenient for our clients, with adequate parking facilities.

Obviously we wish to be as reasonable as possible but we feel that perhaps the only solution is to build modestly on a site where all the above unusual facilities are available.

Yours faithfully
Dacre, Son & Hartley"

The Town Clerk agreed that Rule 5 should apply, the cost of building a new mosque was settled and a fine building was erected in Marlborough Street, the only trouble was the direction of Mecca, but Leeds University were called in to advise on this.

CHAPTER SIX

Natural justice

In the mid sixties I found myself Chairman of the Keighley Rugby League Club, a professional Rugby side attracting gates of 5,000 which amounted to 10% of the town's population. One Saturday afternoon Keighley were playing at Batley on a field at the top of a hill with a distinct slope in the field and inappropriately called Mount Pleasant.

The Batley supporters were superb if you happened to be a Batley player but they naturally did not show the same enthusiasm if you were a visitor. The day I recall was no exception, both Keighley and Batley were vigorously fighting it out, the referee being Tom Watkinson, a schoolmaster from Manchester. There were many penalties to each side and not surprisingly Mr Watkinson decided enough was enough when a Batley player lay flat on his back and he sent off a Keighley player Jack Holmes, who he considered had struck the Batley player. After the game Jack Holmes was indignant at being sent off; he said he had hit no one, (which they usually say). I told him not to worry as he would receive a copy of the referee's report on the incident, the matter would be dealt with by the League Disciplinary Committee and he could either write or make a personal appearance before the Committee if he wished.

Surprisingly on the following Monday the Disciplinary Committee met and suspended Holmes for two matches. He had also not received the referee's report stating why he had been sent off. The next step was to lodge an appeal in the League Appeal Committee. Clearly Holmes had not been told what he had done wrong and the Disciplinary Committee should not have dealt with the matter so speedily. Holmes made a personal appearance before the Appeals Committee and he took with him the player who he was alleged by the referee, to have hit. The player denied that Holmes had knocked him out and the dismissal was a clear case of the referee having wrongly identified the offending player.

The Appeals Committee decided in the circumstances that as there was doubt in the Disciplinary Committee's decision the matter should be referred back to them to reconsider. The Keighley Club wrote to the secretary of the League, Bill Fallowfield pointing out that the Disciplinary Committee having found Holmes guilty, and the Appeals Committee having decided that it was a flawed decision, could not refer the matter back to be re-heard and therefore Holmes should be acquitted.

Bill Fallowfield refused to accept the Keighley view and decided that the matter be referred back to another Disciplinary Committee comprising of three new members who were not on the previous Committee and if Holmes did not attend the meeting it would be decided in his absence. The Keighley Club did not like the Fallowfield attitude and on the morning of the Disciplinary Committee meeting in Leeds the Club and Holmes applied in the High Court in London and obtained an injunction to stop the matter being re-heard. We thought that would have been the end of the matter but not so, Fallowfield persuaded the League to fight on, he was not going to have his authority challenged. The Case came before Mr Justice Danquet in the High Court in London and I attended.

The Keighley Counsel at first did not impress me, he seemed to be drawing out the Case, but then suddenly he began attacking the League in general and Fallowfield in particular. His timing was excellent for he sat down leaving Mr Forrester Paton acting for the League, thirty minutes before the luncheon break. Forrester Paton said if the Keighley Club had acted in a proper way the Case would never have been brought. This brought an interruption from the Judge who remarked that if the Rugby League and especially the secretary of the League, had acted in a proper way the Case would never have been brought. The lunch break arrived and I was well satisfied.

Returning after lunch for the afternoon session Forrester Paton rose to continue and commenced by saying "My Lord, I do not think I impressed you with my arguments this morning, but I am prepared to continue if you think I can persuade you to change the view which I think you have already arrived at". The Judge replied, "Mr Forrester Paton, you can continue as long as you like but I do not think you can persuade me to change my view which you think I have already arrived at". Mr Forrester Paton then withdrew, and we applied for and received costs.

In 1968 the Great Britain team toured Australia and New Zealand, Bill Fallowfield was the Manager and I went along as Chairman of the Rugby League Council. We were together for five-and-a-half weeks during which time we got on well and the Keighley Case was not mentioned by either of us. I have been told that the Case has appeared in the Law Society Exam dealing with an individual's rights.

CHAPTER SEVEN

East Hunslet Liberal Club

I was brought into the picture quite late. The Club Accountant had heard somewhere that a Club could possibly claim the cost of rebuilding a new club under a Compulsory Purchase and he was partly right. I met the Club Committee and explained the legal position.

I was concerned as to the condition of the Club building and also that the Club could be considered as on its last legs, whether there be a Compulsory Purchase or not. I explained to the Committee that to receive the benefit of Rule 5 there were four points to consider and went through them at length.

The Leeds Town Clerk was a man called Potts who opposed my view and made an offer to the Club of £8,000, against my claim under Rule 5 of £100,000 and the matter was referred to the Lands Tribunal. The Club solicitor was Miss Eileen Broadbent who in her younger days had played hockey for England. The barrister was a man called Clough, a first class man who looked very much like Jeremy Thorpe, a one time leader of the Liberal Party. Mr Clough had come to my notice as the man who drafted the Methodist Church Act which went before, and was passed by Parliament. I was surprised when he told me he was a Roman Catholic and I wondered if the Methodists knew this fact.

I was concerned that the Corporation would make a big issue of the dilapidated Club building so I suggested that in order to help the Lands Tribunal we should agree the cost of putting right the dilapidations. Not only did the Corporation agree but to my relief settled an offer of a few thousand pounds, not big enough to seriously affect the claim.

Knowing the Tribunal would make an inspection of the building, I advised the Club that for the period of the hearing they should have as many beer barrels, both full and empty ones, in the cellar to support the contention that the Club had a good trade. The next advice to the Club was to arrange both an old people's trip to the seaside and a children's party, both functions to be prominently displayed on the Club's notice board for the Tribunal's Inspection. The functions could later be cancelled if there was not enough support. A favourable point we could stress was that Keith Miller the District Valuer had recently agreed Rule 5 for the West Bowling Club in Bradford. It would be a bit difficult to explain why a Club in Bradford was entitled to Rule 5, but not available to a Club in Leeds.

In cases such as this it is at times possible to pick up the confidence of the opposition by their actions. The Corporation never called the District Valuer to give evidence on their behalf and Mr Clough hammered this point home by telling the Tribunal that the one man who could have helped had not been called by the Council. "Did this mean that the District Valuer considered the

Council were wrong not to accept that Rule 5 was the correct basis?"

I gave my opinion as to Rule 5 at £100,000 as against the £8,000 offered by the Corporation, agreeing that the Club needed refurbishing but this was not a big cost, a figure had been agreed with the Council and could be deducted from the claim. In my opinion it certainly did not affect the basis of Rule 5. The Council could not challenge this as fortunately they had agreed that figure. I did not have too much to worry about in my cross-examination from the Council.

The next witness was a Club Officer. The night prior to the hearing Mr Clough and Eileen Broadbent had a meeting with the Club Officers and myself. Mr Clough explained how he was going to conduct the Case telling those present that he would need to prove that the Club, but for Compulsory Purchase Order, would be able to carry on and in fact would have done so. He made it clear what he hoped and also made it clear that only the truth should be told and he did not expect anyone to say anything that they did not believe to be the truth. It is surprising how often things do not go to plan. Mr Clough asked the Club Officer one or two questions then sat down. Counsel for the Corporation then started his cross examination of the Club Officer. "Did he not agree that the Club was in such a state that but for the Corporation wanting the premise it would close down?" The Club Officer did not disagree, he also said he agreed that the decoration and structure did need attention. Mr Clough turned round and passed me a note saying he was going to recall me to fill the holes that the Club Officer was digging. This he did. I said I disagreed with the Club Officer's evidence. The Tribunal would only have to look in the cellar to see the volume of trade, the dilapidation was not serious as the amount had been agreed by the Corporation. Probably the reason for the neglect of the Club was due to the fact that it had been known for years that the Corporation were to buy the premises and it would have been a waste of money.

Three months after the Hearing I went to the High Court in London to hear the award read in open court. The Tribunal took the points separately:

1 The premises were used as a Club as claimed.
Agreed

2 There was a bona fide intention to reinstate.
Agreed

3 The premises were of a character that did not change hands in the open market.
Agreed

The award then went on to say that having cleared these three points the Tribunal had to consider whether or not Rule 5 should apply. This was probably the most difficult decision to make but having seen on the notice board at the Club the charitable work such as the old peoples' trips and children's parties, functions usually associated with churches and charities, it had been decided that Rule 5 should apply and duly awarded compensation of £98,000 plus all costs on the High Court Scale.

The following day Mr Potts the Town Clerk and I were invited by Yorkshire Television to appear on the Calendar Programme to discuss the Case. The Case was widely reported and I understand has become a leading Rule 5 Case both in examination questions and Court Cases.

I did for a number of years receive requests for my views regarding Rule 5. One local authority outside Yorkshire had a Children's Home at the seaside in another county. It was being acquired and they asked for my views on Rule 5. Another Case was an unqualified surveyor in the south of England who surprisingly I think, had never heard of Rule 5.

CHAPTER EIGHT

Swine Lane, Keighley

The Swine Lane Case goes back to 1986 when the Upper Airedale Local Plan was placed on deposit, the whole site with the exception of that part which was a former hospital site, was allocated as land with a presumption against development. The area was shown as 54 acres. As part of this appraisal the whole site was reallocated for housing and alteration was published in 1987. The Plan was considered at a Strategic Inquiry into the Local Plan in 1988. In November 1989 the Inspector recommended that the site be allocated for housing.

It was at this point that five freehold landowners of the 54 acres being:

1 Jack Odgen, The Springs, Swine Lane.

2 Chris Ogden, Glen Esk Farm.

3 George Wilman, Field Head Farm.

4 The Yorkshire Regional Hospital Board.

5 Hallbaron Property Co, Devon.

formed a consortium with a view to obtaining planning permission for housing. This was approved with a condition, that Swine Lane Listed Bridge must be widened, also the lane itself must be widened below the canal to Bradford Road. The Council noted in Section 10, page 19 of the Brief, that the Council may be prepared to use Compulsory Purchase powers.

Having retired as Senior Partner in 1980 at Dacre, Son & Hartley and being stationed at the Keighley Office, I knew nothing of the Swine Lane proposal.

Early in 1992 Keighley Key Builders Ltd, of which I am a non-

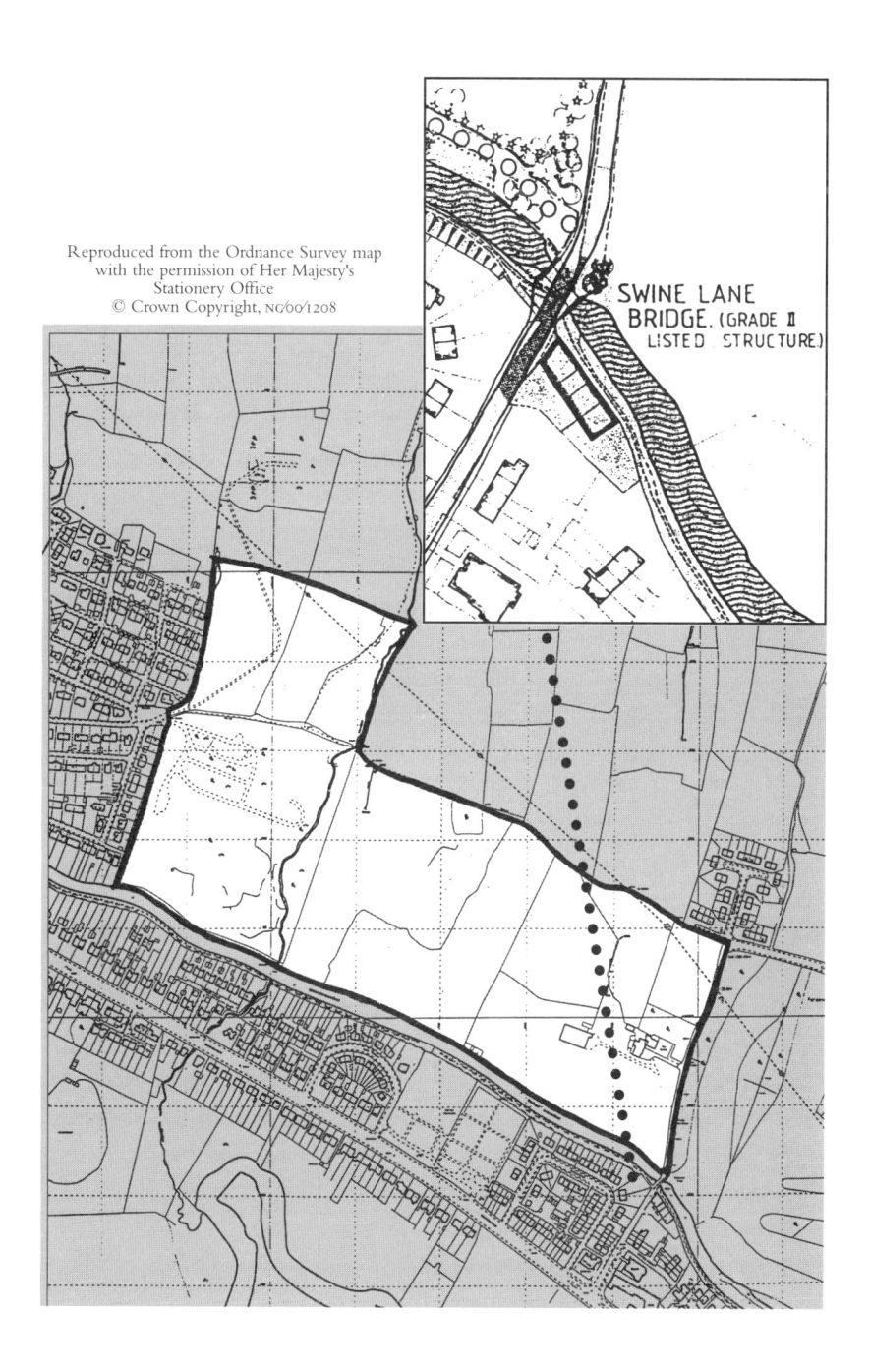

Reproduced from the Ordnance Survey map
with the permission of Her Majesty's
Stationery Office
© Crown Copyright, NC/60/1208

SWINE LANE
BRIDGE. (GRADE II
LISTED STRUCTURE)

executive Director, and my son Andrew Smallwood is Managing Director, sold a barn conversion at Stanbury. The purchaser a Mr West, was having difficulty selling his own bungalow, No 1 Aireview Cottages, Swine Lane and Key Builders agreed to accept it in part exchange, instructing Dacre, Son & Hartley to continue as agents in offering the property for sale. Within a very short time Mr West sent on to Key Builders a letter from Turner & Wall, solicitors enquiring about the ownership of a disused toilet adjacent to No.1 Aireview Cottages. This letter was a surprise and prompted me to pay a visit to the Local Planning Office to discover the reason for it. All was revealed and I realised the effect the planning would have on Swine Lane.

I considered the matter and thought that there could be compensation under Stokes v Cambridge. (This precedent applies when a person has planning permission to develop land but in order to do so, and comply with the terms of the permission, it is necessary to acquire adjoining land. Such land may only be very small in area but if it is essential to the scheme, it is known as 'ransom land' and can result in a vendor's claim for as much as 50% of the total development value).

I wrote to Turner and Wall on 26th August 1992 saying that I considered Key Builders could well have a Stokes v Cambridge claim and I received a reply from them on 3rd September saying they had sent a copy of my letter to Jim Feather at Dacre, Son & Hartley and asked him to contact me.

The only conversation Jim Feather had with me was for a few minutes after a Rotary Club lunch and he told me he had a Counsel's opinion and he had never seen a stronger opinion which was supporting that Stokes v Cambridge did not apply. The next thing was for me to ascertain how many frontagers were affected. The list appeared to be as follows:

Nos. 12, 14, 16, 18, 20 Beauvais Drive
Nos 20, 22, 24, 26 Swine Lane
1 Aireview Cottages, Swine Lane

Aerovac Works, Swine Lane
Ex Methodist Chapel recently sold to Aerovac, Swine Lane

I decided to visit Mr John Eteson, solicitor to Key Builders and was very pleased when he told me he was already acting for the other frontagers. He also informed me that Mr Robert Allen of Messrs Eddisons was acting for all the frontagers with the exception of Aerovac which was represented by Bill Poole of McManus & Poole.

I telephoned Robert Allen and asked him if he had considered Stokes v Cambridge could be appropriate for Swine Lane and was surprised to receive a letter dated 25th September 1992 in which he enclosed three Counsel's opinions obtained by the consortium from Mr Langham. The three Counsel's opinions were dated 27th August 1991, 16th October 1991 and 30th October 1991, all dates prior to the involvement of Key Builders. The opinion of 27th August 1991 read:

> "Finally the consortium will need to bear in mind its position if a Compulsory Purchase Order is made. While the consortium is buying land from the landowners by private treaty, it will be able to control the price paid. Once a Compulsory Purchase Order is made, it will lose this control. The Council will be responsible for paying compensation. The Council already think that it has to pay Ransom Strip Compensation and there will be every incentive for the dispossessed landowners to fight hard for this. If the Council decided to pay such compensation, I cannot see how the consortium could stop it."

Amongst the opinions was a copy of a letter dated 10th October 1991 from the consortium's solicitor to Mr Langham as follows:

> "On the basis of the opinion which the Council has obtained, they will require from the consortium of landowners, a Bond in respect of compensation which they may

have to pay in the event of a Compulsory Purchase Order. In the light of their opinion they will obviously err on the safe side and expect the Bond to be in a sum sufficient to pay compensation on a Ransom Basis. This could be extremely expensive for the consortium and it is hoped that Counsel will be able to give a stronger opinion which may obviate the necessity to contemplate Ransom Level Compensation in the above circumstances."

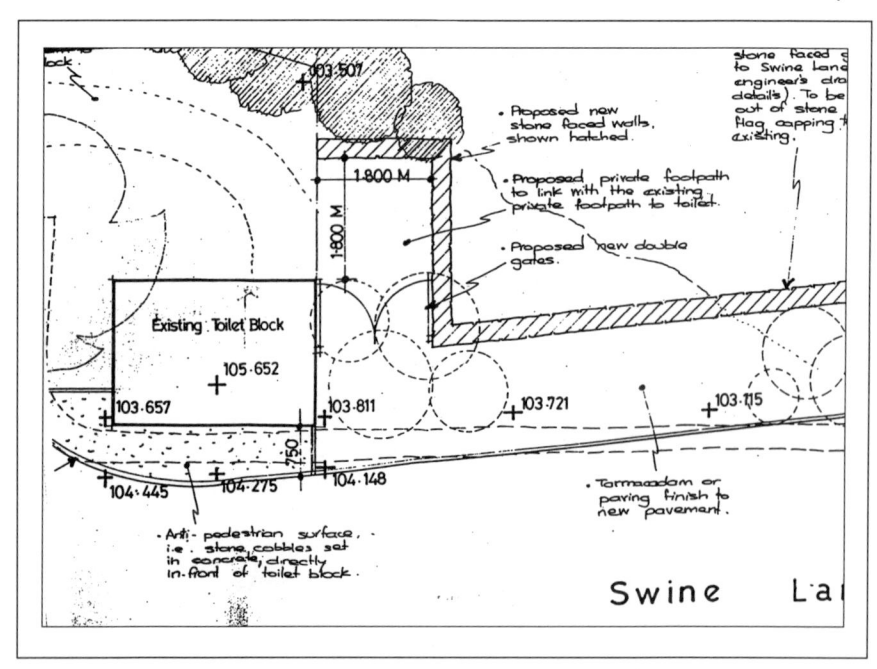

I again visited John Eteson who suggested we obtain Counsel's opinion, but if they agreed, from now on costs should be borne by Aerovac and Key, and if successful they would be compensated for the costs, to be paid by all the frontagers; if not successful Aerovac and Key would stand the cost. This was agreed. John Eteson suggested we should obtain an opinion from Mr Matthew Horton QC who is a leading Compensation Counsel and had successfully won the Ozanne Case at the House of Lords, which in some way was similar to Swine Lane.

It was now the end of 1992 and I was going to Lanzarote for Christmas. Knowing that a retired Superintendent Valuer of the Inland Revenue Valuation Office would be in the adjacent bungalow, I slipped my Swine Lane file into my suitcase and asked him to look at the papers. Three days later he brought my file back to me saying he had read it and expressed the view that I should win. This was some comfort and I looked forward to 1993 when by the time I returned, we should have Matthew Horton's opinion.

CHAPTER NINE

Matthew Horton's opinion

I was about to go on holiday to Lanzarote in 1993 when we received the opinion from Matthew Horton. It surprised me, as he did not think the ransom strip holders had a very strong case, but he did offer to come to Keighley and inspect the site. A date was fixed which meant I had to come home in the middle of my holidays.

A further thing that caused concern to me was that two days before I left for holiday we received a letter from Bradford Council concerning a planning proposal to amend the entry to Aireview Cottages. I went to the Planning Office and saw that the proposal had been made by the Ilkley Office of Dacre, Son & Hartley. It provided for an access across a strip of land in the garden of 26 Swine Lane. The widening of Swine Lane would stop just south of Aireview Cottages and No 1 would no longer have a ransom strip.

Whilst on holiday I thought about Matthew Horton's opinion and came to the conclusion that the plans supplied to him did not show the slope of the land, or difficulty of the canal. When I broke my holiday and came home my son Andrew met me at Manchester Airport and told me that Matthew Horton had postponed his visit to Keighley, but I did not waste my time and was able to deal with the Dacre, Son & Hartley planning proposal. I went to see Mr Eteson and within a very short time the owner of 26 Swine Lane, who wanted to keep my interest in Swine Lane, sold to Key Builders for a nominal figure the site of the proposed new access to Aireview Cottages. This kept Key Builders in the picture. They were still in possession of a ransom strip.

I then telephoned Mr Horsley of Dacre, Son & Hartley, not telling him that Key had agreed to buy the freehold of the ground that his proposed new entrance stood on. I suggested to Mr

Horsley that the proposal was to cut Aireview out of any claim and I further added that the proposal was a departure from, and inferior to, the Town Map. Mr Horsley did not disagree with my remarks and added that the planning application did comply with the Council's minimum requirements and they had reason to think the application would be passed.

Eddisons, who were acting for the majority of ransom strip owners, sent me a copy of a surprising letter received from F M Lister & Son, who had been instructed as surveyors by the consortium. In this letter it made an ultimatum, saying that the consortium's agreement only lasted until 31st March and the Yorkshire Regional Health Authority would have no cash funds available after that date. Therefore a decision from the ransom strip owners must be received by the 25th February 1993. I do not know why the consortium needed money at this stage, they surely were going to receive money, not pay. I think probably the consortium were of the opinion that my appearance on the scene made it probable that Stokes v Cambridge would be vigorously fought, and they wished to put pressure on the other ransom strip owners.

The planning application for the proposed new access was passed which did raise two questions:

1 Why was an inferior plan to that existing, be recommended by the Planning Officer and passed by the Council?

2 Were the Planners helping one council tax-payer against another council tax-payer?

When Mr Horton arrived at the Aerovac offices in Swine Lane we discussed the matter for one hour, during which he said he was not very hopeful of our case. He then said he would like to walk round the boundary of the development land. We accompanied him and the site inspection lasted an hour. We then returned to

Aerovac offices and when we sat round the board-room table Matthew Horton's first words were "I have changed my mind, what I have seen convinces me that Stokes v Cambridge applies". We breathed a sigh of relief.

We then asked Mr Horton:

Q Could the Council make a Compulsory Purchase Order on Swine Lane to help the consortium in their negotiations with the frontagers?

A The answer was no. The Council in such circumstances could not put a CPO on one person's land to help another council tax-payer.

A further question was asked:

Q If the Council put a CPO on the land for widening Swine Lane, quite apart from the development, could we include Hope Value for the estate in our claim against the Council.

A The answer was yes.

Mr Horton said he would write separately on the position.

Mr Horton wrote suggesting:

1 That we should frequently write to the consortium saying that the frontagers were all willing sellers. The only thing was price, but they would negotiate on the basis of a willing buyer and a willing seller.

2 We should not challenge the consortium in contention that the road was quite adequate to take existing and future traffic.

3 Our argument should be the same as the Council that the road must be improved before building starts.

1993 passed without much headway. The consortium just did not want to talk and as far as we were concerned Taywood Homes had not appeared.

CHAPTER TEN

Ransom strips

On 3rd February I received a letter from Robert Allen of Eddisons, saying he had read Matthew Horton's opinion and certainly he was firm that Stokes v Cambridge (50%) applied and furthermore, that the possibility of the consortium achieving a Compulsory Purchase Order was unlikely, if not remote.

On 7th February Robert Allen wrote to Mr Eteson saying the consortium would contract to pay the total sum of £500,000 to be shared by the twelve frontagers and naming them as follows:

1, 22, 24 26, and 26 Swine Lane
1, 12, 14, 16, 18, 20 and 22 Beauvais Drive
Aerovac & Key Builders

This would be subject to the consortium achieving a signed contract with British Waterways in connection with the bridge construction over the canal.

On 22nd March the *Financial Times* reported a sharp rise in land prices because of inflation. In August the interesting case, Wards Construction (Medway) v Barclays Bank was reported and I wrote to Mr Lister who was now acting as the consortium's agent, on 5th August:

"It is now several weeks since we discussed the above matter and I am wondering if the consortium have decided what they wish to do.

It could well be that your clients were awaiting the decision in the Wards Case in Maidstone, if so the result does strengthen Key Builders position. The tribunal decision of £500,000, was appealed against by Kent

County Council - The Court of Appeal referred the matter back to the hands of the tribunal who increased the ransom strip to £2.15 million, probably due to the change in date caused by the West Midland Baptist Case and the increase in building land values since the original award.

Your clients may decide not to sell the land for a while due to possible building land values increasing. I think values could increase, but then of course any one of the ransom strip owners who are prepared to take £45,000 now could have second thoughts, which would be very costly to your clients, should they decide to sell.
To sum up I am getting rather old, I have therefore arranged with Charles Hill of Hill Woolhouse in Leeds, who agrees with my views to deal with this matter should it be delayed a number of years and I am deceased.
Like Barclays Bank in the Ward case, the longer the delay the more valuable the ransom strip becomes, also the same applies to the building land values."

On 22nd August I received a letter from Mr Lister enclosing a valuation he had prepared in January 1992: he showed the Net Development Value at £1,834,966. Ransom Compensation ⅓ at £610,000 against £500,000 referred to in Robert Allen's letter dated 7th February 1994. I replied to Mr Lister on 25th August as follows:

Without Prejudice

"Thank you for your letter of the 22nd August enclosing details of your valuation suggesting figures as to the amounts to be paid to each frontager by your clients the consortium. I am glad that these suggested settlements have been worked out on the Stokes v Cambridge basis, from which I assume and hope that your clients now accept is the correct way forward.

I am personally only interested in the settlement for Keighley Key Builders Limited and Mr & Mrs R Hiley of 26 Swine Lane, Keighley for reasons already explained to you. Mr Poole for whom you refer in your letter, represents the interests of Aerovac.

Working on the Stokes v Cambridge principle all that is necessary is for the development value - in simple terms the difference in the value to a developer and the existing use value - to be established. On page 4 of your notes dealing with your client's meeting with the Council, it was suggested that the value of the land after development costs was £2,295,966 less the value as existing use of £461,000 making a net development value of £1,834,966.

After two-and-a-half years, building land has come out of recession and is now in demand from developers of all sizes. I would suggest that rather than working out hypothetical figures a simple way would be to sell the land, a condition for sale being that the developer would pay all payments under S.106 and other development costs - your clients would then know exactly what they were getting for the land, and by simply deducting existing use value, we would arrive at the true development value.

My view is that on current values and prices achieved for building land the various frontagers on the Stokes v Cambridge principle could achieve an entitlement of anywhere between 1.5 million and 3 million pounds, naturally to be divided equally between them.
I think however that we are getting away from what we

were trying to achieve by bringing the Stokes v Cambridge principle back into focus at the present time. Both Aerovac and my firm, who have stood all the costs so far, have, backed by Matthew Horton's opinion, gone ahead on the basis that we thought that Stokes v Cambridge applies and our common aim in pursuing this matter to the lengths which we have, has been to achieve a substantial settlement for our interests and an equally substantial settlement for the other frontagers.

When however you unexpectedly came forward with an offer of £500,000 for all parties our advisors had to put such an offer to all frontagers. It transpired that the various other frontagers, after being advised that the offer based on Matthew Horton's opinion was inadequate, decided unanimously to go ahead with 'a bird in the hand' approach if a quick settlement could be reached. This would mean all the frontagers dividing the offered amount equally and for the other three parties namely Aerovac, my own company and Hiley to seek an enhanced additional amount satisfactory to our original aims.

It is entirely up to the members of the consortium to decide whether they are prepared to increase their offer in this matter so that a quick and ready solution can be achieved but unless they are prepared to do so then, as you say in your letter, the negotiations will prove to have been abortive. Should that be the case then the consortium will be faced with the alternatives of deferring their proposed sale indefinitely or seeking a CPO through Bradford M.D.C. As Matthew Horton indicated in his opinion he is doubtful whether the Council could obtain a CPO and even if they did one of your own opinions indicates that this would be settled by them on the Stokes v Cambridge principle. Such compensation would have to be paid in advance of any development taking place.

If you think that it would help the consortium by my making myself available for their meeting either with or without advisors I will be happy to so do.

Finally with the limited time constraints you have placed upon me (four working days) it is rather difficult to cover all aspects raised but I have done the best I can in the time to cover most aspects."

After August Taywood Homes appeared and we received a Traffic Assessment from the Tucker Parry Knowles Partnership who were acting for Taywood. The Report covered five points:

1 Existing highway and traffic conditions
2 Development proposals
3 Traffic generated by the proposed development
4 Traffic impact of the proposed development
5 Conclusion

The Report concluded that traffic generated from the proposed development could be satisfactorily accommodated on the adjacent external highway network, subject to various local improvements.

The existing Bradford Road/Swine Lane priority junction and Swine Lane canal bridge would need to the signalised and a new footway would be provided on Swine Lane to the north and across the canal bridge. In addition a footbridge would be provided to connect the development with the southern canal towpath. A new footway would be provided to link the towpath with Bradford Road.

On the south of the canal bridge the footway would join with the existing canal towpath and a footway would be provided to link the towpath with Bradford Road via Beauvais Drive through land controlled by the developer. To overcome existing deficiencies in Swine Lane it was proposed to formulate the shuttle working on the bridge with traffic lights.

In 1992 I had told Key Builders that I considered Stokes v Cambridge applied, but we could expect opposition from the consortium and maybe any future developer. I likened the situation to a small boat crossing the Atlantic in war times. We should be alert and watchful for submarines and should also see we had a good number of depth charges to drop when needed.

Up to the end of 1994 we had only Dacre, Son & Hartley's alternative planning application for the entrance to Aireview to deal with. We had dropped a depth charge on that and retained our ransom strip. Clearly Taywood Homes were interested now. They could have bought, but we were not told. We immediately picked up the view that it was a plan to develop without widening Swine Lane south of the canal. Thus it would not be necessary to buy land from the frontagers and therefore no ransom strips. We had to drop a depth charge on this plan.

CHAPTER ELEVEN

Outline planning application

On 4th January Key Builders received a letter from the Council Planning Officer dealing with an outline application for residential development for land at Leeds/Liverpool Canal, Swine Lane. A notice appeared in the *Keighley News* appropriately dated Friday 13th January 1995 which stated that the applicant was Taywood Homes Ltd. No reference was made to the fact that this plan was a departure from the Approved Development Plan, accordingly it was re-advertised correctly on 20th January.

I was very concerned with the proposal which could be approved. It was suggested that Swine Lane stay as it was, but traffic could be regulated by traffic lights, furthermore the provision for pedestrians could be covered by a separate footbridge. There would be no need to widen Swine Lane south of the canal. I telephoned Dacre, Son & Hartley at Ilkley and spoke to Mr Jim Horsley, discussing the application with him. His comment was that the consortium had reason to think that permission would be granted.

It seemed to me that the consortium position could well have changed and they had reached agreement with Taywood Homes, thus the planning application put in by Taywood Homes. To clear the matter I wrote to Mr F M Lister on 27th January 1995 as follows:

Land at Swine Lane

"I am sorry that I received no reply to my letter of 25th August last, or my follow-up letter dated 21st October.

It would appear that your clients are not ready to buy any of the frontagers' land yet, so I must withdraw any previous offers I have made to you, and should your clients wish to re-open the matter in the future I shall be quite happy to listen to them, but obviously on different terms."

I received a reply dated 1st February 1995:

Land at Swine Lane

"Thank you for your letter of 27th January. I apologise for not writing to you earlier. I did, in fact, go into hospital for a hip replacement operation in October and subsequently had a bit of a complication which necessitated me going back in, so I am afraid my correspondence got neglected at that point.

As I previously indicated to you and to Mr Poole when we had our discussion at Otley, our clients were, at that time, concluding that they could not afford to proceed on the basis of the terms which you, together, put forward.

I have not been invited to attend any further meetings with them since that time, and therefore assume that I have no instructions to try to take the matter further.

If you would like me to ask specifically about the Swine Lane improvement proposals, I shall be pleased to do so, but think that it would be better if I knew exactly where you and Mr Poole stand in the matter, before attempting to call the various landowners together.

I shall be away from the office between 16th February and 23rd but otherwise would be pleased to hear from you."

Having received this letter and as the Planning Application had been put in by Taywood Homes I considered that Taywood would now deal with the matter. The consortium were aware of Stokes v Cambridge. In fact it was covered in their Counsel's opinion in 1991. Taywood could be expected to know all about Stokes v Cambridge and could be expected to put up a fight.

We were concerned that the Council would accept the Taywood application and grant planning permission without widening Swine Lane. We therefore asked WS Atkins, Planning Consultants, for their opinion. They reported on 3rd February 1995 to the Area Planning Officer as follows:

LAND AT SWINE LANE, RIDDLESDEN
APPLICATION NUMBERS 94/03934/OUT
AND 94/03936/OUT

"On behalf of our clients Keighley Key Builders Limited, Aerovac Systems Limited and Fluorocarbon Fabrications Pension Fund, we wish to object to the proposals contained in the above applications.

Our clients fully understand and have accepted the principle of residential development at this location. However, this objection relates not to the principle of development, but to the detail of the associated highway improvements. Both the original (July 1990) and amended (March 1992) Development Briefs for the site required the 'full improvement' of Swine Lane. This would include:

a) the widening of the carriageway to a minimum width of 6.75m between Bradford Road and Carr Lane;

b) the construction of an improved bridge across the canal;

c) the provision of a 1.8m footway along the west side of Swine Lane between Bradford Road and Carr Lane; and

d) the provision of a 1.8m wide footway along the east side of Swine Lane from Bradford Road to Aireview Cottages. These improvements are fully justified on highway grounds and for the safety and amenity of all road users including the existing residents and frontage occupiers. The improvements are required solely to accommodate the additional traffic generated by the development and should therefore be funded by the developer as required by national and local policies.

However we understand from the drawings and from the Traffic Impact Assessment report prepared by Tucker Parry Knowles that these essential highway improvements no longer form part of the developer's proposals. Instead, it is intended to introduce one way 'shuttle' working on the canal bridge controlled by traffic signals. These signals would be linked to other signals at the Bradford Road junction. The intervening length of Swine Lane would not be improved.

Swine Lane is generally rural in character and is typically 5.5m wide. It is a bus route and signed as 'unsuitable for heavy goods vehicles'. Except for a short length on the eastern side, south of Mayville Avenue, there are no footways. Existing pedestrian and vehicle accesses emerge directly onto the carriageway. Some of these accesses have very limited visibility and existing lamp columns, road signs and distribution poles are adjacent to moving traffic.

At the present time we have no basis for challenging the traffic forecasts given in the TPK report. This shows a 50% increase in traffic on Swine Lane as a result of the development. In addition, there will be a significant increase in pedestrian movements. Swine Lane will perform the function of a local distributor road although the existing layout and geometry are wholly inadequate for this purpose.

The proposed traffic signals will result in vehicles braking and accelerating on the gradient approaching the canal bridge. This will increase noise, vibration and atmospheric pollution close to the existing frontage properties.

In the light of the above, the applications would appear to be contrary to policies GP2, H2 and TP7 of the emerging UDP as well as departing from the Development Brief. In addition, a significant increase in traffic in the absence of any improvements is likely to have an adverse effect on the long-term safety record of the road.

It is considered that the full package of improvements to Swine Lane described in the Development Brief are an essential component of the development of this site."

We also picked up the Case of Swan Hill Developments Ltd and British Waterways Board 1995 when it was held that the developers could carry forward their scheme without any special payment to the Board. Interesting but we had more important matters on our minds.

It was reported in the *Telegraph & Argus* on 22nd February that Taywood had two planning applications for consideration, one regarding access and the other for 50 acres, on which to build 400 houses.

The Dacre, Son & Hartley Property Guide for March 1995 stated that 1995 gets off to a flying start, land and new homes boost early figures.

On 24th March 1995 the *Keighley News* reported on the Swine Lane plans. Taywood Homes had withdrawn their planning application when they heard that they faced a refusal on highway grounds.

On 3rd April 1995 we received notice that the Public Inquiry into the Unitary Development Plan was going to be held on 9th May at Shipley. It was abundantly clear that Stokes v Cambridge was going to be decided more on safety grounds than planning.

I was worried at the costs which were being met by Key Builders and Aerovac, if Taywood fought stubbornly they could reach £100,000. I played with figures and thought if 50 acres were being purchased by Taywood, the going rate would probably be £200,000 an acre. On this basis:

Sale price	£10,000,000
Existing value	£100,000
Development value	£9,900,000

I assumed Taywood would have done their sums right and allowed for accommodation works and development value on which ransom owners were claiming their just share.

I endeavoured to obtain from Taywood the price they were giving for the land, but they had declined to give me the information, so I could only assume the price was over £200,000 an acre. Taywood are experienced developers and I would have expected the figure agreed with the consortium to have allowed for the development value and ransom strip values.

We asked Mr Bernard Slater of Cullingworth to provide for us a selection of photographs stressing the importance of road improvements to Swine Lane before any development commenced. These photographs were included in our objection for the benefit of Councillors and the Planning Officer and appear on the following pages.

Just below the bridge, looking towards Bradford Road

Reversing from Aireview cottages into Swine Lane

Examples of danger to pedestrians south of the canal

Traffic dangers on the bridge

Access to Aireview from Swine Lane

Aireview disused toilets, a ransom strip

CHAPTER TWELVE

Plans rejected

On 29th January 1996 the *Telegraph & Argus* reporting on the Unitary Development Plan Public Inquiry which was taking place, stated that the planners had agreed the conditions imposed when outline planning for the site was given in 1992. There had been no advertising or notification of any amendment to the planning brief seen by us, so we were entitled to take it that any agreement between the Council and the developers would insist on the widening of Swine Lane on the eastern side from the canal to Bradford Road. Not to do so, as the Council Highway Engineer so rightly said, would create a development falling a long way short of the requirements of the Development Brief.

On 20th February 1996 the Planning Officer wrote that a planning application had been received for the site. We objected to the proposals, but stated we would have no objection to the proposed development provided that the highway improvements to Swine Lane were carried out prior to the development commencing, as recommended by the Council in its own Development Brief of 1990.

The planing applications were advertised on 9th April 1996. The *Keighley News* reported on 3rd May 1996 that Taywood were appealing to the Department of the Environment for a Public Inquiry because the Council had failed to determine their planning application within the statutory eight weeks. I could not appreciate how Taywood could get over the following:

> On 26th January 1996 a 106 Agreement was signed by the Consortium agreeing to the Planning Brief.
>
> On 29th January 1996 (three days later) the *Telegraph & Argus* reported that the planners had told the Inspector at the Public Inquiry into the Bradford Unitary Development

Plan, that agreement had been reached with the developer who wished to build 400 houses on the Swine Lane site. They had agreed to the conditions imposed in the Outline Planning Permission of 1992. This of course provided for the widening of Swine Lane from the canal bridge to Bradford Road.

On 15th February 1996 (two-and-a-half weeks later) Taywood Homes submitted a planning application ignoring the conditions as to access imposed in the Planning Brief.

We received from Wood Frampton acting for Taywood, a copy of a letter dated 3rd October to Mr Morris at the Planning Office referring to a meeting they had had the previous day and asking that their planning application be reported for determination at the Planning Committee on 14th November 1996. I was worried about this letter. Could it be interpreted as meaning at their meeting on 2nd October Mr Morris had agreed Taywood proposals in principal and was prepared to recommend acceptance?

April Chamberlain (née Brett) a young solicitor in the office, was helping John Eteson and one of her duties was to attend Council committee meetings, sit in the public gallery and supply us with a verbatim report of the meetings. She was brilliant with her reports. At the meeting on 14th November the Committee considering the matter was the Town & Country Planning (Keighley) Subcommittee.

Mr Morris the Area Planning Officer at Keighley Town Hall introducing the matter, told the Committee that any decision of the Subcommittee that morning was not binding, but was rather a recommendation which would then be passed on to a superior Committee, the Policy and Plans Committee. The duty that morning of the Committee therefore was to make a recommendation to the Policy and Plans Committee. This Committee meeting although not binding, would give us a clue as to the view taken by the Planning Officer and possibly the relationship if any, with Taywood Homes.

Mr Morris gave a report of the planning history of the site and mentioned major changes in the revised proposals for a new canal crossing linking (via a footpath) onto Cliffe Crescent and onto Bradford Road. He then referred to the number of objectors and named them, mentioning the letter written by Mr Eteson on behalf of the frontagers. He stated that the decision before the Subcommittee was whether they would recommend approval of the amended application. The main questions to be addressed were those of traffic and pedestrian concern. We could not disagree with Mr Morris so far. He then went on to say the new proposals suggested that Swine Lane bridge should remain unaltered (although traffic lights were to be installed to ensure a through flow of traffic). Swine Lane was to be improved northwards to the corner of Carr Lane but no footpath was to be provided; instead pedestrians were to be discouraged to use Carr Lane onto the canal bridge down Cliffe Crescent to the bus stop and shops on Bradford Road.

The Planning Officer then recommended the Subcommittee should approve the amended plans, subject to a satisfactory S106 Agreement. Several people spoke against the amended plan and points were made:

> 400 houses means approximately 800 more cars. Swine Lane had to be widened and proper footpaths supplied.

Councillor Slater recommended that the Policies and Plans Committee refuse the application on the following grounds:

1 Highway provisions

2 Pedestrian safety

He also commented:

> The Section 106 Agreement was legally binding. This contained a requirement that Swine Lane bridge should be widened in keeping with the Development Brief drawn up with the full knowledge of the recommendations of the Highway Engineer.
>
> The Highway Engineer was of the opinion that the proposed traffic measures were inadequate.
>
> In terms of planning gain, little was being offered. The gain lies with the developer in not widening Swine Lane. They would not have to pay out compensation, a seven figure sum.

A vote was taken and it was held by a majority that the Sub-committee would recommend that the application be refused.

A letter dated 2nd December 1996 was received informing us that the Public Inquiry would be held on Tuesday 4th February 1997 and that the Council would argue that the proposed development:

> a) Does not accord with the Development Brief prepared for Swine Lane and referred to as existing supplementary planning guidance to continue in force in Appendix C to the deposited Unitary Development Plan.
>
> b) Would generate an increase in vehicular and pedestrian traffic along Swine Lane which has no continuous footways and more particularly on the canal bridge, which was severely substandard and would lead to conditions detrimental to the safety of pedestrians.
>
> c) was unsuitable and would be detrimental to the personal safety of users of the footbridge.

On 3rd December 1996 the Committee met to consider the Report to the Town & Country Planning Subcommittee and they recommended to grant planning permission subject to conditions and a Section 106 Agreement. It was reported to the Committee that no works were proposed to Swine Lane between the canal bridge and Bradford Road. It further stated that there would be a new canal bridge for pedestrians linking via a new footway down along Cliffe Crescent on to Bradford Road.

It did not mention pedestrians visiting houses facing on to Swine Lane. There were a number of objections that had been received under highway implications including:

1 Traffic light system was inadequate and Swine Lane bridge should be widened to take extra flow of traffic.

2 Increased volume of traffic signals would not be able to cater, thus leading to vehicles backing on to Bradford Road.

3 Pedestrian facilities:
Inadequate width of the footway over Swine Lane bridge extremely hazardous for pedestrians. Pedestrian link over the canal was inadequate. Concern was expressed about the lack of a pedestrian facility along the whole length of Swine Lane to Bradford Road.

The Highway Engineer was of the opinion that the proposed traffic manager measures were not suitable for such a large development. The increase in vehicular movements along Swine Lane could be detrimental to road safety in the absence of any major improvements and realignment. He also expressed concern about the inadequate provision of pedestrian facilities along Swine Lane, which could lead to conditions detrimental to the safety of pedestrians with a potential for increasing pedestrian movements generated by the development.

The footbridge across the canal was also of concern. The steep gradient would be unsuitable for wheelchair users and it was not designed in accordance with the advice set out in DB32, which recommended that footpaths should be intervisible.

Conclusion and Recommendation

"The proposed scheme does not accord with the Development Brief in terms of the highway works, which were identified as being required by the Design Brief and shown on the UDP proposals.

The issue therefore is quite clearly whether as an alternative, this scheme can safely accommodate the existing and additional traffic and pedestrians likely to be experienced on Swine Lane.

The question of pedestrian flow is more questionable: There is an existing low level of pedestrian movement down Swine Lane from Carr Lane to Bradford Road. The design of the scheme is anticipated to intercept the movement and channel it through the development site and over the new bridge. This scheme will solely accommodate all the likely pedestrian flow within the site, and to schools, shops and buses in a manner which as far as practical, avoids the use of Swine Lane itself as a pedestrian facility. Given the low level pedestrian movement which exists on Swine Lane at the moment, then it is conceivable that such a scheme could be supported and would be a reasonable alternative.

It must therefore be quite clearly seen that the choice facing the Council is whether this revised scheme (with the retention of 13 acres of woodland), is materially more beneficial and in planning terms, more acceptable than the outline permission which has been granted, and which requires the demolition and replacement of the listed bridge at Swine Lane.

Whilst in prime highway terms, the approved improvements to Swine Lane does handle predicted flows adequately, it does mean there would have to be possibly in terms of the previous resolution of C & E Committee in April 1992, compulsory purchase of properties fronting Swine Lane in order to achieve widening.

Recommendation

"That planning permission be granted to the development subject to an Agreement under Section 106."

It was to the Planning Officer's credit that he did report to the Committee all the points for and against the Appeal in a fair way. It was his recommendation I found strange. Naturally I was concerned, to me and I am sure to many others, the application for the revised scheme was to avoid paying a large sum of money to the Ransom Strip owners.

For the planners to suggest a planning permission be granted subject to a Section 106 Agreement was indeed a surprise. Surely a 106 Agreement should be in place before granting the permission. Once the permission is granted there would be scope for arguing that there were no ransom strips.

There were points to take into account such as the opinion of Matthew Horton QC that the Council could not make a Compulsory Purchase Order to benefit one council tax-payer against another council tax-payer.

I did worry at the way things were going, one usually does. There was a Public Inquiry by a Ministry Inspector to deal with the whole Swine Lane matter on 4th February 1997. I could understand Taywood trying to beat the Inquiry by getting a decision quickly in their favour, but surely a Planning Officer should wait until he had the benefit of the Inspector's report which would be held six weeks later?

I had not long to wait. I was telephoned on the evening of Friday 20th December to be informed that the Town & Country Planning (Policies & Plans) Subcommittee had finished their meeting and rejected the Taywood Plan supported by the Planning Officer. They had been unanimous in their decision.

Although it was a relief to receive this decision we had not taken it for granted and had prepared for a different verdict. If the Committee had taken the advice of the Planning Officer and passed the application it would have been a departure from the Town Map and the Minister would have had to give his approval. This in such cases, is merely rubber stamping.

We had taken the precaution of contacting Garry Waller, the Member of Parliament explaining the position and asking him to request the Minister to call in the planning application by Taywood, to be decided by the Minister himself. Within a matter of hours we were contacted by the Department of the Environment who invited us to meet the Director of Planning from the DOE North, in Leeds. We made it clear to the DOE Director there was no evidence of misconduct by anybody, but we considered this planning application was of a nature that the Minister should call it in to be decided by him. The DOE Director listened to us fairly, asking us questions, then informed us that he did not think he could advise the Minister to call it in, but promised that he would watch the application very carefully. I think if the planning application had been passed it would not have been rubber stamped, but fortunately it was not.

John Eteson and April Brett had handled the job well. I could go to sleep and anticipate a happy Christmas and prepare for the Public Inquiry of 4th February 1997. I did not think Taywood would have a happy Christmas.

CHAPTER THIRTEEN

Public Inquiry

We realised that if Taywood had been successful in their planning application on 20th December the Public Inquiry would not have been held. Taywood with some justification, would have been able to claim that they had a planning permission without widening Swine Lane and development would not be held up by having to buy ransom strips.

We would have been able to challenge the Town and County Planning (Policies & Plans) Subcommittee if they had given permission as it would have been contrary to the Unitary Plan. The Minister would have to give his approval, though it is usually rubber stamped, but we would not have given in. The Inspector at the Public Inquiry at the Unitary Plan had been told that the developer had signed a 106 Agreement to carry out the improvements to Swine Lane before development started. The 106 Agreement included widening Swine Lane. I could see a costly battle ahead. We were planning for the Public Inquiry on 4th February and costly though it would be, we had to put out the best team we could. Naturally we wondered how the Council would deal with the matter.

Matthew Horton QC was not available on the day of the Inquiry and on John Eteson's advice, we instructed John Taylor QC, a Counsel in the same Chambers as Matthew Horton. The Brief was sent to Mr Taylor on 17th January. On the 27th January, John Eteson, April Brett, Mr Smith from Aerovac and I went to London to meet Mr Taylor when we discussed the case with him. We were impressed with Mr Taylor; he was unusual but just the man to handle our case, we spoke the same language.

I had already said the appeals should be thrown out on pedestrian safety alone, but Mr Taylor said the Taywood Plans were of equal importance and we must produce the best expert witnesses to challenge the whole of Taywood's proposals. The planning was as important as the pedestrian safety.

On the evening of the 3rd February Mr Taylor arrived from London, staying at Oakwood Hall Bingley. The Inquiry was expected to run for four days at the Town Hall, Keighley. Those who are familiar with the Council Chamber will know it is oblong in shape, the Chairman sitting on a raised dais at one end and fixed seating arranged in a horseshoe around it. I arrived at the Town Hall very early, realising that the Taywood team who were appealing, would sit on the Inspector's right and the Council team on the Inspector's left. Before any of the others came I claimed the seats facing the dais so that Mr Taylor had a table to place his papers on immediately in front of the Inspector. It was a very opportune move because it was a full house and we had secured for Mr Taylor a prime position. The Inspector was Mr J J Parkinson MSc CEng MICE MIHT MRTPI.

Taywood had a planning permission to carry out the development in accordance with a valid 106 Agreement, which stipulated that they had to widen Swine Lane. The reason for the Appeal was to avoid widening Swine Lane and so avoiding having the ransom strips owners demanding large amounts. Many people and organisations objected to the Taywood Appeal suggesting that the land was not suitable to develop as planned and they should not be permitted to develop the site. To revoke planning permission would have been very costly as the freeholders would have been able to claim compensation based on the loss of development value from the Council.

People simply appealing against the planning had no hope of revoking the planning permission already given, and although no doubt getting personal satisfaction, was costing the frontagers dearly. They no doubt felt very hurt at the proposed development, speaking at length with no effect and extending the length of the Appeal over four days. Mr Taylor's agreed fee was £15,000 for Day One and £1,500 for each following day. Taywood as the appellants were first to speak. They were represented by Mr Purchase QC. It was quite clear to me that Taywood had assumed that they would obtain the planning permission they had asked for at the Committee Meeting on 20th December. The failure to secure it had resulted in their having to put a team together quickly. Mr Purchase was brilliant and presenting the Taywood Case he called the following witnesses:

1 Mr P Frampton BSc(Hons) TP ARICS MRTPI
   Partner with Wood Frampton

2 Mr D Tucker MSc CEng MICE MIHT
   Partner with Tucker Parry Khowles.

3 Mr N Lowe BA Dip Arch RIBA
   Partner in the firm Birkett Cole Lowe

4 Miss K Reginni
   Employed in the firm of Countryside Planning and Management.

5 Mr J Cocking F Arbor A FRES
   Consultant to Popperwell Associates

The Council was represented by Mr A Crean, Counsel from Manchester. He cross-examined the Taywood witnesses and we were well satisfied with him. He called to give evidence:

1 Mr C Wagget BA MSc MRTPI
   Senior Planning Officer

2 Mr B Hunt IEng AMICE, Senior Engineer

3 Mr W Caulfield DipTP, Team Leader Transportation and Planning Division

4 Mr J Gorman BEng, Principal Development and Buildings Officer

The Council team performed well dealing with cross-questioning from Mr Purchase and presented a good case.

Mr Taylor cross-examined the Taywood witnesses and was brilliant. He reminded me of the television programme "Rumpole of the Bailey".

Appearances and interested parties:

1 *The Frontagers*
   Mr J Taylor QC

He called:

    Dr R Wools BArch PhD Dip Cons RIBA
   Principal of Roger Wools and Associates

    Mr C Dallas BSc MIHT
   Employed in the firm of Bryan G Hall

    Mr I Tavendale F Arbor, Arboricultural Consultant

2 *NHS Executive Northern and Yorkshire Region*
   Mr C Brook FRTPI
   Managing Director of Clive Brook Associates Ltd

3 Mr B Whitaker, Local Resident

4 Ms A Dent, Local Resident

5 Miss J Brown, Local Resident

6 *Bradford Urban Wildlife Group*
   Mr L Barnett, President
   Mr H Firman, Member

7 *Aire Valley Conservation Society*
   Mrs P Ward, Secretary
   Mr Bowler, Member

8 Councillor M Leathley, Ward Member and Deputy Chair of the Local Planning Committee

9 *Riddlesden Action Group*
  Mr C Pepper

10 *Bradford Ornithological Group*
  Mr Radcliffe, Chairman

11 *Morton Village Society*
  Sir Anthony Tippett, Committee Member
  Mr A Plumbe BSc

12 Councillor S MacPherson BA(Hons)

I was well satisfied the way the Inquiry was going and was amused when one of those attending, a typical Keighley man, came up to me after listening to Mr Taylor and remarked "I wish yon fellow had acted for me last year when they caught me driving when I was drunk and lost my licence. I would have got off with that man defending me."

The four days put aside for the Inquiry was not sufficient for the Inspector to hear all the evidence put forward by the various interested parties and the Public Inquiry was therefore adjourned to Monday 9th June and was expected to finish on Tuesday 10th June. The two days were again not sufficient and there was a further arrangement to reconvene on 14th and 15th July with a site inspection on 16th July.

We considered we had three very good witnesses and they served us well.

Dr Roger Wools

A member of the Royal Institute of British Architects, B Architecture (1st Class Hons), Diploma in Conservation Studies and Doctor of Philosophy. Architectural experience has included projects for English Heritage, The National Trust, The Civic Trust, National Museum of Science and Industry's Railway Museum at York and National Museum of Wales. Also managed the re-listing of the buildings of

England and Humberside and is currently chairing a working party into the future use of redundant chapels in Wales. Also a consultant to the Council of Europe.

Mr Christopher P Dallas
A member of the Institution of Highways and Transportation and a graduate member of the Institution of Civil Engineers, BSc Civil Eng. Local Authority experience previously gained with Kirklees Metropolitan Council and Dept of Engineering, and the London Borough of Harrow involving the design and construction of highway improvements, traffic management and development control. Currently involved in traffic related matters in commercial residential and industrial developments including major town redevelopment schemes.

Mr Ian Tavendale
An independent arboricultural consultant operating throughout the North of England, elected fellow of the Aboricultural Association, January 1989. Member of International Society of Aboriculture and Royal Forestry of England, Wales and Northern Ireland and consulting Arboriculturist for Mortgage and Insurance Users Group.

We considered Mr Taylor's closing speech was good, bringing out all the valid points clearly and we were reasonably happy.

Closing speech of John Taylor QC on behalf of Keighley Key Builders Ltd, Aerovac Systems (Keighley) Ltd and Others

PUBLIC INQUIRY. KEIGHLEY TOWN HALL 15th July 1997

General
In this case the starting point must be that there exists for the Appeal site a valid planning consent capable of implementation

recently granted in January 1996 and the subject of a valid Section 106 Agreement. It is not a case where a consent is needed to allow development to proceed, development can proceed and there is no fetter upon it.

There is required, of course, the purchase of other land for the roadworks but the frontagers, whom I represent who are an interested party to this extent, have always made it quite clear that they are as a body willing to sell by agreement. The City Council have always indicated in any event that compulsory purchase powers would be used if needed. The fact that land is required from the frontagers is not, of course, a surprise to the owners or their developers. They have known that since 1989 when the Public Inquiry was told of the Greater Airedale Structure Plan that "it was accepted" that Swine Lane would need improvement. It was no surprise to them as this has been known for eight years.

The essence, therefore, of the present Appeal proposals has nothing to do with the sudden tenderness for the canal bridge, nor is it justified by safety considerations rather the reverse; in my submission it is an attempt to avoid the consequences of the Plan for the release of this land which has been known since 1989 and indeed through the Local Plan process, solely for financial reasons.

Mr Frampton in his evidence referred to sums required by the frontagers and jeopardy to the development and delay. It is not, however, the frontagers who delayed gaining Planning Consent from 1992 to 1996; nor they who have been unwilling to treat.

They remain so. What is however interesting in this matter is that despite Mr Frampton's assertions, the appellants have called no evidence to support the alleged jeopardy to the scheme showing that the scheme is unable to bear the costs of acquisition of the land in Swine Lane and to the works so that the scheme becomes economically unviable. The point is like the Emperor without clothes in the fairy story – it is naked and without substance. Certainly it is not made, as it could have been, by putting development appraisals to the Inquiry, upon which the proposition could be tested. The frontagers are well aware of the promise of the City Council to use compulsory powers if needed and have no reason not to cooperate. Any monies payable on the acquisition are, of course, dependant on the economic viability of the scheme; they do not make it economically unviable - that is a matter of common sense.

I said at the outset that these are not matters, however, relevant to the determination of this Appeal. They do not assist you. The central question here is whether these appellants should be allowed to set aside a fundamental condition of the development which was laid down in 1989, namely the provision of a modern, suitable and safe access.

One other preliminary point is worthy of note. The improvement of Swine Lane, south of the bridge is not in itself linked to whether the bridge remains as it is or is rebuilt. It is really a separate issue of safety, linked with the bridge, but existing in its own right, since, keeping the bridge unaltered, in order to improve the situation, does not carry the consequence of justifying no works to improve safety south of the bridge. They may continue to be needed even if the bridge is kept unaltered and it is the frontagers case that that is so. The developers have sought in this hearing to link the two, but keeping the bridge unaltered does not itself lead to the conclusion that no works are required on safety grounds south of as far as Bradford Road.

The Development Plan
In examining the proposal it is important to bear in mind the history of the emerging Development Plan and the past events surrounding the allocation of this land. These events took place of course within the full application and context of circular 8/87 which sets out guidance which is still in substance applicable to listed buildings.

The fundamental point upon which discussion of the release of the Appeal site was predicated has always been the creation of a new acceptable and safe access to modern standards for both vehicular traffic and pedestrians at Swine Lane.

The Development Plan starting point was the Greater Airedale Strategic Plan, in which the site was allocated for development. The Inspector's Report into the objections to that plan so far as the site was concerned was dated 6th November 1989 and you have it. He made clear in that report that he had given weight in allocating sites which were "sensitive" to the authority's intentions (paragraph 11 of the letter). Among these expressed intentions was the intent to prepare Development Briefs and that the release of the site would require improvement of Swine Lane, including bridging the canal (paragraph 1.10.3). It is important to note that the objectors (ie: the owners) accepted that fact and acknowledged that the potential development was adequate to fund the necessary works. The improvement of Swine Lane was endorsed by the Inspector in paragraph 1.10.12 and he recommended that work to be inserted in the Transport Policy. The Inquiry had full discussion of the need for a "new bridge" (reported in paragraph 1.10.3) and the remarks of the Inspector (reported in paragraph 1.10.12) have to be viewed in that context, possibly his envisaging additional bridge with retention of the old.

In January 1990, however, the Enterprise and Environmental Committee of the City Council resolved that the recommendations be considered in the context of the UDP and that a Brief be prepared for the site. The first Brief emerged in 1990 and provided in paragraph 4.1 that access should be solely from Swine Lane and set out in detail the proposal to widen and improve Swine Lane including a rebuilding of the existing bridge. That Brief also included a proposal for a footbridge over the canal. An Amended Development Brief emerged in March 1992, which reiterated the proposals for Swine Lane, but omitted the footbridge because of the blocking of the footpaths.

There was thus complete consistency of approach. The release of the Appeal site at all stages of the formulation of development was to be based on the improvement to modern standards of Swine Lane, including a new or rebuilt bridge over the canal.

In the draft UDP in 1993 the Appeal site was shown as a committed Housing Site that reflected a resolution (to which I have already referred in 1992) to grant Planning Consent subject to a Section 106 Agreement. The site was subject to objections and discussed at the Inquiry. I ask you to note that in the evidence of the City Council the authority stressed the need to gain suitable access from Swine Lane in accordance with the Brief and on certain conditions. A Highway statement was forwarded to the Inquiry setting out the required improvements to Swine Lane and noting that the 1994 applications were undetermined and showed inadequate improvements. So the position was made quite clear at the Inquiry and it was open to the owners to come and make objections – but they decided to stand by for tactical reasons. No objections were made by them or by Taywood Homes.

There were no objections made to the Plan by the owners of the land or Taywood, nor any suggestion that the Plan as conceived on the basis of the Brief should be set aside, or in any way modified. The Plan was allowed to progress, therefore, on the basis first set out in 1989 and confirmed in the Development Briefs and no attempt was made to refute the Traffic Statement in the evidence of the Council. The Traffic Statement made by the Council was not contradicted by the owners.

It was on this basis that the Plan was confirmed by the Inspector allocating the site for housing and containing in the proposals report a reference to the Swine Lane Scheme and the Development Brief to inform the public of the access proposals. The references are at page 34 and page 35 of the Keighley Report and page 141 of the General Part. The Swine Lane proposal is incorporated by reference in the justification which is part of the Plan. It is defined on the proposals map. That report was accepted by the City Council and no modification as proposed to Policy TP7 with its reference to Swine Lane and the Development Brief.

It is to be noted that the proposal reports comprise "written statements providing additional details and justifications of the UDP proposals shown on the proposals maps" (page 141). They are thus part of the Development Plan. Page 129 also refers to the Swine Lane Brief and stipulates that it is to remain in force (appendix C).

Government Policy places great weight on the plan-led system and attaches importance to the role of the public in the formulation of the Plan. The UDP has now reached the final stage in the process. A fundamental issue arises in this Appeal as to whether the Inspector stands by that position or not. As I have said before, the owners and the Development Company have had ample chance during the Plan process to object to the proposals incorporating the Brief, the reference to the Brief and to the access arrangements. At no time have they done so, but, stood by and allowed the process to move forward. Members of the public who may have wished to object to the allocation on the basis that the Swine Lane proposal was being altered and no modern access provided have had no chance of that. This application is in substance an attempt to modify the allocation in the Plan at the stage when the Plan is about to be adopted without going through the Plan process. It is an attempt at a time when the Plan is on the brink of adoption and one which should not be countenanced if the plan-led system is to have any meaning.

The proposals are contrary to the Plan and to the proposal's report in particular for this site, which incorporates the Development Brief by reference and great weight should be given to that. While the Section 54A presumption will not arise until adoption, the proper approach, having regard to Government policy and Law, is to give the Plan full weight and approach the matter on the basis that the Plan is to be applied until material considerations indicate otherwise.

Of course stress has been placed on the Officer's Report recommending approval but that was rejected by the Council, who adhered consistently to the position on the basis of which the scheme was promoted. In this respect they were clearly more in line with the principles underpinning the plan-led system than the Officers.

Other material considerations
There are three other material considerations in my submission:

1 Whether there is justification to retain the bridge unaltered on Listed Building grounds overriding considerations of finding a safe, suitable, modern access;

2 Whether the traffic arrangements based on retention of the bridge unaltered are sufficiently safe and appropriate;

3 Whether the footbridge proposal as part of the scheme is acceptable in the Conservation Area.

As to number 1, it is wholly self evident that a modern housing development of 400 houses needs a safe appropriate access to the main road system. The developer's case is basically that the self evident proposition should be disregarded because it requires the late 18th century bridge over the canal to be substantially altered, or demolished (whichever word you use, I do not think it matters particularly).

Clearly, it must be accepted that Listed buildings should be preserved unless there is some justification according with policy which permits that policy to be set aside. It is also trite law that "special regard" has to be paid to the interests of preserving Listed Buildings. PPG15 sets out the proper approach on this matter.

My learned friend always likes to spangle his cases with odd trumpery of legal glitter and we were referred to *Heatherington* and *Bath Society* in opening and are to have *Shimuzu (UK) Limited* in closing. In my experience, this trumpery glitter rarely adds much to the argument that is not resolved by common sense and the fact of the case being dealt with. All that *Heatherington* tells us (page 234), for the purposes of this Inquiry, is that if Section 54 points to a different result from the result obtainable by having "special regard to the desirability of preserving" a Listed building it is for the decision maker to weigh the matter and arrive at a Judgment. *Bath Society*, a Conservation Area case, merely states the obvious to the decision maker, that where special attention needs to be paid to the conservation or enhancement of a Conservation Area, it is entitled to considerable importance and weight (page 670). At the end of the day it is still a question of balance.

Shimuzu (UK) Limited (1997) concerned an arcane compensation issue relating to the nature of works done to Qantas House. No compensation was payable for demolition of part of a Listed building but if the works were an alteration of the building, compensation was payable. This is a nice semantic issue. The House of Lords reversed the Court of Appeal on the basis that whether works constituted demolition of a building or only an alteration was a question of fact in each case to be resolved by the decision maker. Demolition involved a substantial clearing of the site amounting to total destruction, works for partial demolition were an alteration to the building.

Applying that principle to this case, for what it is worth, it is a mere matter of words, how the works are described. They are either a significant demolition of part and retention of part or an alteration involving a significant demolition but retaining part. The important point is not these semantics but how acceptable is what is proposed to be done.

So far as the points about the works to the listed building are concerned, it is important to bear in mind that the whole matter was visited in 1992; the Listed building application went to the Secretary of State who did not call in the application for Listed Building Consent, despite the representations of English Heritage. There is therefore already a judgment in place to the effect that the need for a new safe access for the 400 houses on the released land justifies the works to be done. That position has not altered.

That judgment is supported by Dr Wools with a distinguished background in work for the Heritage. The Appellants on the other hand, called no acknowledged Listed buildings expert to support their case. Evidence was given by Mr Lowe, who, whatever thoughts he may have had about the bridge, clearly had very few thoughts initially about the Conservation Area as far as the footbridge was concerned. It is the submission of the objectors that Dr Wools' evidence is to be preferred. He did not undervalue the bridge, but clearly, under cross-examination, stood by his view that its essential integrity was sufficiently maintained if works were done on the lines of the Listed Building Consent, the only obvious difference in appearance being the longer tunnel as viewed from below and the footpath, and the clearer vision for drivers on the road. The local form of construction will still be there to experience in exferance retained western facade together with the reformed western face and the bridge form from the towpath. This is not a case of total removal but a justified change to secure an important planning objective, namely a safe access to the new development in a context which is safe for existing residents.

Preservation of the bridge does not in any event justify an unsafe road access being retained. There are still the problems to which I have already alluded to the south of the bridge in terms of pedestrian safety, which need resolution, which can be solved independently by widening. I ask you to bear this solution in mind. The better solution clearly, however, is to provide proper access as envisaged in accordance with the approved solution and the objectors believe that can be done without losing irretrievably the interest and quality of the bridge. The City Council's compromise was right in 1992 and is still right, and should be supported as balancing properly the development plan consideration on the one hand, and the Listed building considerations on the other.

The second consideration is that of safety. The prime issue in this case is whether the traffic proposals are safe and acceptable. Mr Tucker comes with a novel proposition that a 400 house new development can be permitted which takes access from a clearly substandard road leading to the main distributor. The case was first started on the basis that the pedestrian traffic would virtually all be removed from Swine Lane where there was no footpath, and given traffic control and in the absence of pedestrian conflict the access would be safe. Even he realised that that position was untenable, and in his revised information accepts that pedestrian traffic (existing and new) would remain on Swine Lane, while the daily car volumes rose from 4363 to 6523. He says 47 existing and new, compared with 53 now on the bridge, and 50 south of the bridge compared with 68 now. These figures are only judgmental and depend on how adequate are the alternative footpath arrangements within the site and over the footbridge.

There can of course firstly be no doubt that the present arrangements are substandard, although they do not merit improvement given scarce highway resources. Under the proposals they remain substandard. Much was made by Mr Tucker of the absolute width of the road, giving space for pedestrians and traffic flows. The effective widths tell another story, given that there will be prams, wheel chairs, and cyclists, who on Swine Lane south cannot squeeze into the "one foot" available by the wall to avoid the traffic. They are in it and given only 11 seconds in the red phase to skelter across the road from side to side of the highway. The crossing pedestrian who does not stand in the place appointed by him to cross by Mr Tucker (of course who will not be there to police it) could face cars requiring stopping distances for more than the 24 metres of visibility available particularly where the driver speeds up to beat the lights confident that no car is coming in the other direction. The added flow will worsen the environment for pedestrians and the queues advancing in platoons be more polluting and more intimidating - the road flow well exceeding Buchanan's Environmental Level in peak hour.

The stark point is that Mr Tucker advocates perpetuation of substandard conditions for an increased vehicular flow and says for pedestrians that there will be slightly fewer of them and that it is justifiable. Even if he were right in his figures of pedestrian movement, a decision maker must be robust indeed to accept that the road will be safe, to predicate that accidents will be no more than currently when 400 houses use this road as their main access, imposing on women and children the need to walk in a narrow carriageway with increased flow.

In fact Mr Tucker's assessment of pedestrian movements the objectors believe is wholly fallacious. He assumes (i) that the safe pedestrian route is overwhelmingly attractive and (ii) that bus services in the new development will obviate the need for people to walk to buses on the Bradford Road. Mr Dallas sets out for the existing journeys the likely change within a range between 17 and 47 continuing to use the bridge, and 57 to 74 south of Swine Lane giving averages of 32 and 65 respectively. Mr Hunt's view that after development the flows could be 103 and 117. The notion that the new development, given Figure 1 volumes of trips in Mr Tucker's New Information by walking, would generate on Mr Dallas's average flow from 400 houses no more than 15 people extra (Mr Tucker's 47) and south of Swine Lane a reduction of 16 (Mr Tucker's 50) is frankly unbelievable.

The route over the footbridge is circuitous, for many parts of the estate no shorter than using Swine Lane, it is not user friendly with changes of levels and a bridge and route through trees and past places which by walkers at dusk (as early as 4 pm in winter) and at night will be perceived as quite unsafe. Based on the character of the scheme, Mr Tucker's view is implausible, an implausibility enhanced by his assumption that the 20 minute 727 bus service will be uniformly preferred to the 10 minute more rapid service in Bradford Road.

Mr Tucker's position is an extreme position. It is comparable to his position on every other issue - the visibility line over the bridge of 33.2 metres south is the best one for his case, the width of the road (despite poles and other obstructions) is taken as the absolute rather than the effective width. It is not a balanced approach. Future pedestrian flows are essentially matters of subjective opinion and it is unsafe not to take a more realistic view than his of the matter. Common sense alone suggests that with an increase of 400 dwellings in the locality and a choice of route the flow will not fall but rise and Mr Hunt's more practical view should be given in my submission, great weight.

One reflection has occurred to me as the Inquiry has lengthened whether in looking at this matter the participants of this Inquiry have not allowed themselves to be influenced unacceptably by the appellant's thesis that if a fall in pedestrian flows occurred the addition of over 2,000 vehicles per day on this wholly inadequate road is to be regarded as safe and acceptable. That is a strange proposition when looked at in the cold light of dawn. The starting point is that however many pedestrians there are walking, increased traffic, however controlled on the bridge, and Swine Lane south, all are intermingled on an inadequate carriageway - not just on sunny days but cold winter dark night peak hours - children coming from school, cyclists, women, joggers, and shoppers and that is from any view hazardous and unsafe. It needs correction. To disregard that and expose future users of this road from the existing and new development whatever their numbers to increased flows is not an acceptable approach in modern planning terms. To use the bridge as an excuse to turn common sense considerations of safety on their head and ignore the risk to pedestrians abrogates responsibility for insuring that people are protected from increased risk to each one of them (even if fewer) of personal injury or loss of life. There can be no justification for giving planning consent for this application to plan new development with an absolutely unsafe access which you are asked to do. Even if numbers are smaller there in an increased risk to life and limb.

So far as the bus point is concerned, the idea that all residents will use an improved 20 minute 727 service avoiding the walk to Bradford Road is of course part of Mr Tucker's thesis, to reduce pedestrian flows on Swine Lane. Predicting people's habits and preferences for the future is notoriously difficult and Mr Dallas had the more sceptical approach that some will continue to use the Bradford 10 minute buses and that to my mind is not unreasonable. Buses are missed, services perceived to be more rapid and frequent are preferred at both ends of the journey. Provision of buses will not eliminate either present or future bus users from walking down Swine Lane and the use of any new service should be seen in that practical light.

Whether the service will be viable after five years is highly problematic. The assumptions made by Mr Tucker are very sensitive to small changes on trip rates, modal split and cost. The 727 is presently a subsidised service - given the illustrative past schemes for detached houses, the site location and character for employment and schools, high rates of bus users, characteristic of inner urban areas, is unlikely. The relevance of this uncertainty is only this - that of Mr Tucker's case on pedestrian flows lasts for the full term of the future of the development. The subsidies are only for five years. There is no basis for safe prediction that bus services will continue in perpetuity and Mr Lomax's letter is wholly silent on these economic questions.

Mr Purchase has several times set the ground for saying that the proposals now are better in PPG13 terms by having public transport provisions for a five year support in the Section 106 Agreement and by implication of course this supports the Appeal. That point is only as good as the viability assessment. A five year subsidy does not underwrite long term bus provision. This "improvement" in the Agreement has no substance unless Mr Tucker's prediction comes true. For the reasons already given, I suggest that acid scepticism about the bus service on this site is likely to be a better approach. In any event the bus consideration must rank small in the context of issues of safety to life and limb.

I turn to the third point. The proposals of the appellants may be designed to preserve the bridge but they clearly do violence to the Conservation Area in relation to which there is a special statutory obligation, namely to ensure its conservation and enhancement. The introduction into this canal landscape of either the first, second or third footbridge neither conserves nor enhances the area. All are sizeable visual intrusions of an urban nature and require the removal and lopping of trees, affecting visually both the area and the cemetery.

To my mind the casual treatment of the footbridge proposals by the City Officers in their report, almost total indifference to it, was a culpable omission in the proper exercise of their powers. I hope you and the Secretary of State will not imitate this given your statutory responsibilities. So little did the appellants give proper weight to this issue that they had a second and even a third go at the footbridge design. I hope that you will approach the matter by looking at the application as it is at this stage and not by taking the view that matters can be sorted out at the detailed plans stage. Nor given it is a proposal in a Conservation Area, does it do for it to be said that this is an outline proposal and the details can be sorted out later.

Mr Tavendale's evidence and Dr Wools' makes it quite plain what is the extent and range of the threat that construction of this bridge poses to the area.

That arises at all stages – not just the final stage – the preparation and construction stage, the effect of the bridge itself and at the maintenance stage as leaf drop on the bridge and other impacts give rise to pressure from users to make the footbridge safer. This morning you heard evidence from Mr Tavendale and Mr Cockin. The general principles are agreed between them but you will have to choose between their evidence on the specific points which are not agreed. Drawing 565/2 of Mr Popplewell addresses in part the future but like the rest of the appellant's case (perhaps he has taken training lessons from Mr Tucker) presents a rosy picture which tells us nothing of what will go on as a result of the effects of construction or of the bridge supports on the trees. A wonderful picture has been painted of sensitive manhandling during construction with the care of a Tree Surgeon being applied. The idea is ludicrous when examined in the cool light of day. The whole picture is unrealistic. This is a major structural engineering work in the context of this area. The more realistic picture, I put it to you, is that put forward by Dr Wools and Mr Tavendale. Mr Tavendale has, of course, considerable experience of supervising such works on site. The evidence is clearly such that no sound judgement can be formed that the footbridge however built will not scar the Conservation Area significantly and the likelihood is that it will. The proposal is neither conservation nor enhancement and its effects need to be weighed against the alleged advantages of preserving the bridge.

Conclusion
In the last analysis this case turns on whether what is proposed represents a development which provides safe and proper access for its residents without detriment to existing local people. Once the issue on that matter is resolved the Conservation considerations which do not given consistent signals, become a matter of easy resolution. It is the objectors case that, in modern planning terms, Swine Lane is not safe for its users with current traffic. The proposals increase the risks from higher traffic flows, probably to even more pedestrians, many of whom will be children. Mr Tucker's ratio approach to risk lacks logic and carries the absurd consequence that the more traffic on the same flow of pedestrians the more the risk diminishes of injury. Mr Dallas' approach on this point is the better one and more in line with common sense. The risk practical experience tells us will be greater and safety less. A new development in the year 1998 needs

a proper safe access. That conclusion is as true now as in 1990 at the time of the Development Brief. It should not be set aside.

An acceptable Listed Building Consent does exist for a replacement to the bridge. It may still need some engineering refinement and redesign in detail to take account of Mr Tucker's very late note delivered last Friday on this matter. The length to be reconstructed of Swine Lane north of the bridge allows ample scope for the tweaking of the vertical alignment. It is the objectors' view that the consent represents a proper balance of the safety and Listed Building considerations. Moreover, preservation of the bridge is at the clear expense of the Conservation Area with the footbridge proposal so that the Conservation issues point in opposite directions.

The Conservation interest in any event does not affect safety considerations as far as Swine Lane south is concerned. Even with the bridge retained, that still needs proper improvement which the Appeal scheme does not provide.

In the end, you are being asked to endorse unsafe road proposals long term which do not accord with the planning policy background for this site, which the owners during its formulation accepted. The attempt on their part to subvert the process which involved public consultation should not be blessed with your recommendation for approval and the objectors seek the dismissal of the Appeal.

We received notice on the 18th November 1997 that the Taywood Appeal had been refused by the Minister.

CHAPTER FOURTEEN

Financial negotiations

We were hopeful that 1998 would bring an early settlement, the matter having gone on for six years and Taywood said that they wanted the matter settled. Having spoken with Mr John Scannell the Taywood Director, we again confirmed our view that the way forward was to take the value of the land and deduct from this the existing use value. The result is the development value on which the frontagers are entitled to their share.

Example:

50 acres with planning permission for 400 houses at £200,000 per acre	£10,000,000
Existing value say	100,000
Development value	£9,900,000

Taywood then produced an agreement that was so one-sided that we dismissed it. I cannot comment on the contents as we were asked to treat the draft agreement in confidence. We repeated to Taywood the suggestion previously made, that if they disagreed with our valuation of the land they should in confidence, tell us what they were paying for it. They refused, which may be because they were paying more than £200,000 per acre. It was interesting to note that my figures were supported by Robert Allen and Bill Poole, both like me Fellows of the Royal Institute of Chartered Surveyors, Valuation Division. Taywood seemed to rely on planners rather than valuers in reaching their decisions.

We received from Taywood a financial appraisal which appeared to accept the figure of £200,000 an acre but then discounted it to allow for accommodation works and what they claimed to be abnormal cost of construction. An original idea but not appropriate. We were again asked to treat their appraisal figures in confidence.

In September John Scannell asked for a meeting which was arranged by John Eteson. Robert Allen for the residential frontagers, Bill Poole for Aerovac and myself for Key Builders were all present. John Scannell said he wished to reach a settlement so that building could start in the New Year. He wished to introduce his financial appraisal for discussion and told us that they had a contract to buy Ogden's land. Mr Scannell seemed to wish to develop Ogden's land as the first phase. He did not seriously object to a figure of £200,000 an acre but said this figure should be reduced as set out in his financial appraisal for accommodation works and cost development of the site, reducing the value to £75,000 an acre. We were not impressed and told him so.

We were shown a layout plan of Ogden's land which we were told was the only land on which a contract of sale was in existence. It was interesting to see that the access roads were provided from Ogden's to Procter's Land. What was the point of this if Procter's land was not going to be developed? No mention was made by John Scannell of the Planning Brief 1992 which had a condition that all the consortium land should be one development and not carried out piecemeal.

A few days after the meeting Mr Scannell telephoned me to enquire what Key Builders would accept, I referred him to John Eteson. John Eteson told me that John Scannell suggested that Taywood would pay the frontagers £1,250,000 to share. I contended this was not enough. If Taywood had paid Ogden £210,000 an acre then it was reasonable for me to say that the Consortium land of 50 acres was worth £10,000,000, the existing value a maximum value of £500,000 leaving a development value of £9,500,000.

Matthew Horton QC who acted in the House of Lords Ozanne Case advised us that the percentage of the development value payable to the ransom strip holders should be 50%. It was no argument for Taywood to say that the cost of carrying out the development was exceedingly high. They should have known the cost of developing the land and allowed for it in the price offered.

Towards the end of the year Taywood made another planning application suggesting that the bridge need not be widened, but a steel footbridge be added to cross the canal. We supported Taywood's proposal provided the carriageway for vehicles was widened. The Inspectors report at the Public Inquiry suggested that the stone listed bridge would be costly to widen. A new pedestrian bridge alongside might be acceptable, provided the carriageway was made safe for the increased vehicular traffic the development would cause.

On 23rd October 1998 the *Keighley News* quoted Councillor Martin Leathley, a member of the Keighley Planning Sub-committee as saying about the development "Time is running out. There is planning permission to build and all the company has to do now is provide a means of access."

The Planning Subcommittee deferred a decision on a steel footbridge over the canal in the hope that Taywood would come up with a better design.

CHAPTER FIFTEEN

The end of the saga

We entered 1999 hoping that the matter would soon be settled, seeming to start every year with this wish, but on 23rd February there was a good sign for Taywood made us an offer, subject to contract and without prejudice of £1,500,000, to be apportioned out amongst the claimants. This amount was to include payment to British Waterways for necessary rights to work under or over the canal.

Before agreeing any payment to British Waterways the Case of Swann Hill Developments Ltd and British Waterways 1995 would need to be looked at. In this case it was held that the developer could carry forward their scheme without any special payment to the Board and the offer also was conditional on each party bearing its own costs. At last there was hope that an early settlement could be reached. The Taywood offer was insufficient, but at last it was an offer. I had, since the negotiations began, realised that I was very vulnerable, firstly being eighty-four and secondly being a non-executive director of Key Builders. The other frontagers could say I had a conflict of interest. I also realised that I was very unpopular with the consortium who refused to let their surveyor talk to me initially when they probably considered that my appearance was going to be costly. The time had arrived for my involvement to be transferred to another party. I thought about the matter and considered that the most suitable person would be Charles Hill, a partner in Hill Woolhouse of Leeds and he agreed to take over from me.

As soon as Charles Hill had read my files he thought it would help if Taywood Homes brought in new blood and accordingly John Carter, Regional Managing Director started direct discussions with Charles Hill. The position has not been easy. As Matthew Horton QC had told us, the Council could not place a

Compulsory Purchase Order on the ransom strips as to do so would be favouring one council tax-payer against another. As applied in this case, where the Ransom Strip owners were willing sellers the only thing to agree was the price. The position could be prejudiced if any of the following situations occurred:

1 The consortium refused to sell their land. Anyone could refuse though not likely as obviously they would appreciate the money from the sale.

2 Taywood Homes could buy the land from the consortium then put the land in the Land Bank and negotiate with the frontagers at their leisure.

3 Any one of the frontagers could refuse to sell his land although not likely, because the price would take care of this, but the position could be influenced if one of the Conservation Societies who were opposed to the scheme, bought one ransom strip from one of the frontagers, then decided not to sell it to the developer.

The Parties were not likely to disagree except on price and it was in the interests of all consortium, frontagers and developers to reach agreement. In June a provisional offer was agreed. Taywood instructed Walker Morris, solicitor of Leeds, to act on their behalf and at last we were making progress, though not without complications. British Waterways instructed Gerald Eve & Co, Chartered Surveyors, to advise them and they requested four to six weeks to make their report.

In September agreement was sent by Walker Morris to John Eteson, who fortunately acted for all the frontagers. The Contract was as expected, complicated and all frontagers were requested to sign. This was not easy for John Eteson as a few frontagers started raising fresh problems, but it was eventually completed. Taywood expressed a wish to complete the ransom strip sale by the end of

October as they wished to enter the land to start the accommodation works.

Looking back it was in 1987 that the consortium land was changed from Green Belt to housing. My involvement did not start until 1992 when we wished to negotiate for Key Builders on a ransom strip basis. It has been a hard fight; the frontagers team of solicitors, barristers, surveyors, town planners and other professionals were excellent and could not be bettered and my thoughts are that Key Builders and Aerovac should be acknowledged by the claimants for their involvement in paying all the costs exceeding £100,000, on the basis of the amount only being recovered if successful; if losing the frontagers would not be liable to pay Key or Aerovac one penny. Had funds not been made available by the two firms the outcome could have been very different. In 1997 John Taylor QC, representing the frontagers at the Public Inquiry privately told me he rated our chances of winning at 60% and I was comforted as the Case proceeded that he gradually increased the odds.

All the parties, consortium, frontagers and Taywood Homes have gained experience throughout the Case and will remember Swine Lane for a few years to come.

I thought the matter was finished and the papers could be filed away but it was not to be. So many people had been concerned and just a few picked up a six figure amount for a narrow strip of garden for road widening. Locally it became news, rather like 'Who wants to be a millionaire'. Naturally the local weekly papers and the Bradford evening paper picked up the story, but it came as a surprise when Yorkshire Television requested a live interview with John Eteson to discuss the matter in their evening Calendar Programme. This was followed by half a page in the *Yorkshire Post* and a full page in the *Daily Mail* which considered the matter warranted sending a helicopter to take aerial photographs of the site. *The Times* published a 12 in column reporting the matter and named both John Eteson and myself. The result being it brought letters, some of which were from people who were asking for advice and prepared to pay us for giving it. At eighty-four, as I had officially retired over twenty years ago, it was quite a surprise.

Looking back over eight years one thing stands out, the longer the settlement took the higher the claim became. In 1992 the ransom strips could probably have been acquired for between £50-100,000 and in 1999 the claim was agreed at £1,652,500 which meant together with the costs of the Developer, the total would probably amount to about £2,000,000.

CHAPTER SIXTEEN

Memories

I often think of not only a surveyor's life pre-war but also the changes that have come about within all professions. When I became an articled clerk with Hollis & Webb in Leeds at seventeen I shared an office with a junior partner sitting on a stool at a plan table while he sat at a roll-top oak desk. Office dress was a suit except on Saturday mornings when flannels and a jacket were allowed.

All staff including articled clerks were called Mr, Mrs or Miss, Christian names were not used by anyone. I signed my articled agreement, agreeing that I would take the Chartered Surveyor examination, (that was before it became the Royal Institution), also I would take the Chartered Auctioneer exams. The agreement also stipulated that when I left I would not practice for a given number of years within five miles of Leeds Parish Church. When I joined my brother and Douglas Hartley on the formation of Dacre, Son & Hartley I was still not twenty-one and very much still learning. Hollis & Webb were chartered surveyors not estate agents, they sold property by auction but surprisingly passed private treaty sales to other firms. When I found myself at Dacres in Skipton in the beginning 90% of my work was residential sales but I worked on the commercial side which grew considerably. I had to provide my own car and purchased a Flying Standard 12 hp for £205. Petrol in those days cost one shilling a gallon, the same price as a packet of cigarettes which seems through life to have remained constant, a pack of twenty equalling a gallon of petrol.

Pre-war the self-appointed London experts had not reached Yorkshire which did help local firms such as Dacres, it also created a trusting and closer relationship between clients and surveyors. Quite a number of surveyor's advertisements today would not have been acceptable by the Institution. The Surveyor today has much to worry about. Planning requirements are much more severe and his work on Compensation Law is much more complex due to the Lands Tribunal decisions. I was fortunate because after the Thwaite Chapel CPO which I referred to in Chapter Four, I was asked to join the Property Division of the Methodist Church based in Manchester. The Methodists quite rightly relied upon the individual Trustees of local chapels to manage their own affairs, but on sales they had to obtain confirmation from the Property Division and it was this requirement I found interesting because quite often I was asked for a free opinion on the advice of a local agent. This enabled me to obtain information on what was going on in important and unusual cases, some of which were later to became leading cases on Compensation Law.

At Dacre, Son & Hartley's Keighley and Skipton offices I was responsible for the professional staff who were trained within the firm and taking their examinations. No one was ever employed who came to Dacres from another firm. A further credit which I am proud of is that in forty-four years no one was made redundant or sacked. I had a case where a young boy helped himself to £70 out of the tenant's rent till. I did not dismiss the boy, but gave him a lecture on dishonesty and told him I was not going to sack him but instead gave the lad one month to find another job. If he was still with the firm at the end of that time I would then find a reason to dismiss him. He duly left and three years later I met him in the street and asked how he was doing. He told me that after he left Dacres he had gone to a firm of accountants and was awaiting his examinations' results. Today I would be looked upon as being very much out of date and I think the same thing applied to all of my 'vintage'.

CHAPTER SEVENTEEN

Other professions

Over the years all the professions have altered dramatically. I opened my first bank account in 1936 with £100 and was given a bank book showing my transaction written in ink. From the lovely banking hall with oak counters and no bullet-proof screens I was invited into the manager's office for a personal introduction which was conducted over an oak rolled-top desk and after ten minutes I was offered a cup of tea. As a twenty year old I was most impressed.

Naturally with the nature of my work I came into contact with all the local bank managers and excellent men they were, living in the town and taking a major part in the activities and social services. Sadly over the years due to computerisation and the internet, banks have changed. They will no doubt claim that they have moved with the times but many clients miss being able to go into their branch and have a friendly chat with the manager. The present corporate manager who looks after two or three branches, as well as making personal calls at the homes of customers, is not able to give the personal service that once was offered. The general policy of banking today must make the corporate manager's job a very lonely one, as the full offices of staff which once processed the daily business are no longer occupied.

Banks are nowadays more profit conscious to the detriment of the customers and the recent public outcry regarding the charging for use of cash machines is surely justified when these were initially installed to cut the cost of staffing counters. Whilst progress is inevitable it is not always in the interests of customer confidence and relationship.

Alongside banking there have always been people advertising financial services, offering to look after your money. I have often thought if these people are as good as they say, why do they not look after their own money instead of other peoples', it would be much more profitable to them. I occasionally read in the papers notices of receivership and it is surprising that financial advisers are sometimes given as the occupation of failure. I think that there are three ways to lose money:

1 The quickest – horse racing
2 The nicest – women
3 The most sure – financial services

My father often used to say "He that goes a borrowing goes a sorrowing". I amended the version by adding "unless you can pay it back on demand".

There are still some good people looking after peoples' money but these should be appointed by recommendation rather than the self advertising we receive almost daily through the post.

From when I became an articled clerk in a chartered surveyor's office I realised how closely the two professions of solicitor and surveyor worked together in relation to buildings and land, both had a strict code of conduct as to what could or could not be done. Solicitors could not advertise how good they were, their goodwill depended on their record given by satisfied clients. Now solicitors are apparently allowed to advertise even on television, offering services to people who have been involved in accidents and they will handle the claim free of charge. It is assumed they then take a fee out of the winnings.

I have always had a good relationship with solicitors as they contributed greatly to my goodwill. When I first started in Keighley I only knew one firm but I was eager to have all the local solicitors' work. I decided to make a Will with each firm and afterwards tore them all up, except one. Only one firm actually sent me a bill for three guineas but it was money well spent for that firm instructed me to sell a farm by auction in Harrogate a few weeks later.

My work brought me in contact with many solicitors and with a few exceptions they were a pleasure to work with and especially in days long past, very professional and easy to form a friendship with.

At eighty-five I am sure people will think I am very old. In fact a few months back two men stopped me in Keighley town centre. One of the men enquired if I was John Smallwood. When I confirmed that I was he remarked that he had thought I was dead. We talked for a couple of minutes and as we parted, after going about five yards I heard one of the men say to the other, "He's failing you know". I was able to shout after them, "Yes, I am, but I can still hear you!".

I think the secret of old age is that when you retire you must work not for money but to keep your mind active. That is what I try to do.